Approaches to Research on the Systematics of Fish-Borne Trematodes

Approaches to Research on the Systematics of Fish-Borne Trematodes

Jitra Waikagul

Urusa Thaenkham

ELSEVIER

AMSTERDAM • BOSTON • HEIDELBERG • LONDON
NEW YORK • OXFORD • PARIS • SAN DIEGO
SAN FRANCISCO • SINGAPORE • SYDNEY • TOKYO

Academic Press is an imprint of Elsevier

Academic Press is an imprint of Elsevier
32 Jamestown Road, London NW1 7BY, UK
The Boulevard, Langford Lane, Kidlington, Oxford, OX5 1GB, UK
Radarweg 29, PO Box 211, 1000 AE Amsterdam, The Netherlands
225 Wyman Street, Waltham, MA 02451, USA
525 B Street, Suite 1900, San Diego, CA 92101-4495, USA

First published 2014

Notices
Knowledge and best practice in this field are constantly changing. As new research and
experience broaden our understanding, changes in research methods, professional practices,
or medical treatment may become necessary.

Practitioners and researchers must always rely on their own experience and knowledge in
evaluating and using any information, methods, compounds, or experiments described herein.
In using such information or methods they should be mindful of their own safety and the safety
of others, including parties for whom they have a professional responsibility.

To the fullest extent of the law, neither the Publisher nor the authors, contributors, or editors,
assume any liability for any injury and/or damage to persons or property as a matter of products
liability, negligence or otherwise, or from any use or operation of any methods, products,
instructions, or ideas contained in the material herein.

British Library Cataloguing in Publication Data
A catalogue record for this book is available from the British Library

Library of Congress Cataloging-in-Publication Data

A catalog record for this book is available from the Library of Congress

ISBN: 978-0-12-407720-1

For information on all Academic Press publications
visit our website at store.elsevier.com

This book has been manufactured using Print On Demand technology. Each copy is produced to
order and is limited to black ink. The online version of this book will show color figures where
appropriate.

Working together
to grow libraries in
developing countries

www.elsevier.com • www.bookaid.org

CONTENTS

PREFACE

"Approaches to research on the systematics of fish-borne trematodes" is aimed to serve as a guide book for the student in Parasitology, Biology, or Molecular Biology, or for any young researcher who is interested in studying fish-borne trematodes, in particular on their morphological identification and genetic makeup.

In-depth knowledge of medically important fish-borne trematodes is well known in their endemic regions, but some coexistent species are not known or have been reported as the better-known species. Eggs of small liver and minute intestinal flukes are similar in morphology and sizes, causing difficulty in microscopic diagnosis. On the other hand, many species of minute intestinal flukes exist in the same regions, and some of these concomitant species are not known to laboratory workers and may be reported as the well-known species in the area. Hence the present knowledge may not reflect the true situation: in particular, the prevalence of major species of fish-borne trematodes—small liver flukes may be overreported, while intestinal flukes may be underreported. It is important to realize the gap and to work to produce new knowledge to bridge the gap existing at present.

The authors would like to thank members of the Faculty of Tropical Medicine, Mahidol University—Tippayarat Yoonuan for the photographs, and Wijak Anowannaphan and Akkarin Poodeepiyasawat for their technical assistance on the illustrations.

Jitra Waikagul
Urusa Thaenkham
Bangkok
December 2013

CHAPTER 1

Medically Important Fish-Borne Zoonotic Trematodes

The trematode or fluke—a flatworm in the phylum Platyhelminthes—has a dorsoventrally flattened, leaf-like body. The medically important species belong to the subclass Digenea, the endoparasites of verte-brates. A digenean trematode has a complicated life cycle, which com-prises adult and several developing stages: egg, miracidium, sporocyst, redia, cercaria, and metacercaria. They utilize a minimum of two hosts (blood flukes). The final or definitive host for adults and the snail's first intermediate host occurs in the stage cercaria, which is an infective stage. For other flukes, their life cycles include a second intermediate host that contaminates with the parasite's infective stage: the metacer-caria. The genus *Alaria* of family Diplostomatidae has an additional developing stage, mesocercaria, which is a stage between the cercaria and metacercaria stages.[1]

The trematodes utilize fish as their second intermediate host; the so-called fish-borne trematodes comprise about 12 families: Acanthocolpidae Lühe, 1909; Bucephalidae Poche, 1907; Clinostomatidae Luhe, 1901; Cryptogonimidae Ciurea, 1933; Cyathocotylidae Poche, 1926; Diplostomatidae Poirier, 1886; Echinostomatidae Looss, 1902; Heterophyidae Odhner, 1914; Opisthorchiidae Braun, 1901; Psilostomatidae Odhner, 1913; Strigeidae Railliet, 1919; and Troglotrematidae Odhner, 1914. However, not all families are known to infect humans; those fish-borne trematodes that have been reported in man are in 5 families: Clinostomatidae, Echinostomidae, Heterophyidae, Opisthorchiidae, and Troglotrematidae. Among those that infect the human species, the opisthorchid fluke has the most pub-lic health importance; it has been recognized as a type I carcinogen, and chronic infection by this liver fluke leads to cholangiocarcinoma development. The heterophyid intestinal fluke sometimes coexists in the endemic region of the liver fluke and can cause confusion in diag-nosis and prevalence since eggs of both the opisthorchid and hetero-phyid flukes are similar. Nanophyetid infections have been reported in

Approaches to Research on the Systematics of Fish-Borne Trematodes. DOI: http://dx.doi.org/10.1016/B978-0-12-407720-1.00001-7

North America and eastern Siberia. Clinostome and fish-borne echinostome are considered minor in man.

1.1 MAJOR FISH-BORNE TREMATODES

1.1.1 Opisthorchiidae

The liver fluke in the family Opisthorchiidae comprises 4 genera distributed in America, Asia, and Europe (Table 1.1, Fig. 1.1). Infections by the following species have been reported in man: *Amphimerus pseudofelineus* (Ward, 1901) Barker, 1911, and *Metorchis conjunctus* (Cobbold, 1860) in America; *Clonorchis sinensis* (Cobbold, 1875) Looss, 1907, *Metorchis orientalis* (Tanabe, 1921), *Opisthorchis noverca* (Braun, 1903), and *O. viverrini* (Poirier, 1886) Stiles & Hassall, 1896, in Asia; and *Metorchis bilis* (Braun, 1893), *C. sinensis*, and *O. felineus* (Rivolta, 1884) Blanchard, 1895, in Europe.

Amphimerus is known in North America as a cat liver fluke and has been reported to cause bile duct fibrosis and pancreatitis in cats. Human infection with this fluke was first mentioned in Guayaquil, Ecuador, and was referred to as a new species: *Opisthorchis guayaquilensis* (Rodriguez et al., 1949).[57] *O. guayaquilensis* was later considered a synonym of *O. pseudofelineus* Ward, 1901, and based on its morphologic distribution of vitelline follicles, which separate into anterior and posterior groups with the latter extending to the testes level, it was moved to the genus *Amphimerus*.[58]

Recently, the infection in humans was confirmed as *Amphimerus* by the morphology of adult worms and electron microscopy scans of eggs.[46] A total of 23.9% (71/297) stool samples positive for eggs was observed in the native people (Chachi) of the northern coastal area of Ecuador. The positive egg rates of the 3 investigated villages were 15.5, 26.7, and 34.1% with the highest rate found in the most remote village 120 km from the coast. According to Calvopina et al,[46] the identity of *Amphimerus* in Ecuador is still unsettled; whether it is *A. guayaquilensis* or *A. pseudofelineus* requires further study. The livers of three cats and three dogs from one of the villages, were all positive with many adult worms of *Amphimerus*; this liver fluke is a zoonotic pathogen of domestic animals and humans in the area. Infections by the *Amphimerus* fluke in domestic animals have been reported in Central and North America, suggesting the possibility of human

Table 1.1 Geographical Distribution of Fish-Borne Zoonotic Trematodes

Parasite	Country
Clinostomatidae	
Clinostomum complanatum	India,[2] Israel,[3] Japan,[4] Korea,[5] Thailand[6]
Echinostomatidae	
Echinochasmus fujianensis	China[7]
Echinochasmus japonicus	China,[7] Lao PDR,[7] Japan,[7] Korea,[7] Thailand,[7] Vietnam[7]
Echinochasmus liliputanus	China[7]
Echinochasmus perfoliatus	China,[8] Japan[7]
Echinoparyphium paraulum	China[8]
Echinoparyphium recurvatum	China,[8] Egypt,[8] Indonesia,[8] Taiwan[8]
Echinostoma cinetorchis	China,[8] Indonesia,[8] Japan,[8] Korea,[8] Taiwan[8]
Echinostoma hortense	China,[8] Japan,[8] Korea[8]
Heterophyidae	
Apophallus donicus	Oregon-USA[9]
Ascocotyle longa	Brazil[10]
Centrocestus armatus	Japan,[11] Korea[12]
Centrocestus caninus	China,[41] Taiwan,[8] Thailand[13]
Centrocestus cuspidatus	Egypt,[14] Taiwan[15]
Centrocestus formosanus	China,[8] Lao PDR,[16] Philippines,[8] Taiwan,[17]
Centrocestus longus	China,[8] Taiwan[18]
Cryptocotyle lingua	Greenland[19]
Haplorchis pumilio	China,[41] Egypt,[20] Iran,[8] Lao PDR,[21] Philippines,[22] Taiwan,[17] Thailand,[23] Vietnam[24]
Haplorchis taichui	Bangladesh,[8] China,[8] Iran,[8] Lao PDR,[21] Pakistan,[25] Philippines,[26] Taiwan,[17] Thailand,[27] Vietnam[24]
Haplorchis vanissimus	Philippines[28]
Haplorchis yokogawai	China,[8] Indonesia,[29] Lao PDR,[30] Philippines,[31] Taiwan,[32] Thailand,[27] Vietnam[24]
Heterophyes dispar	Korea[8]
Heterophyes heterophyes	Egypt,[33] Iran,[41] Israel,[34] Tunisia,[8] Turkey[35]
Heterophyes kutsuradai	Japan[36]

(Continued)

Table 1.1 (Continued)

Parasite	Country
Heterophyes nocens	Japan,[37] Korea[12]
Heterophyopsis continua	China,[8] Japan,[18] Korea[12]
Metagonimus minutus	China,[12] Taiwan[38]
Metagonimus niyatai	Japan,[39] Korea[39]
Metagonimus takahashii	Korea[12]
Metagonimus yokogawai	China,[40] East Indies,[40] Indonesia,[8] Iran,[41] Japan,[40] Korea,[12] Philippines,[40] Taiwan,[40] Rumania,[40] Russia,[8] Spain[40] Ukraine[40]
Procerovum calderoni	Philippines [22]
Procerovum varium	Japan[41]
Pygidiopsis summa	Japan,[42] Korea,[12]
Stellantchasmus falcatus	Hawaii,[43] Japan,[44] Korea,[8] Philippines,[22] Taiwan,[32] Thailand,[39] Vietnam,[24]
Stictodora fuscata	Japan,[45] Korea[12]
Opisthorchiidae	
Amphimerus sp.	Ecuador[46]
Clonorchis sinensis	China,[8] Japan,[8] Korea,[8] Russia,[47] Thailand,[48] North Vietnam[8]
Metorchis bilis	Russia[49]
Metorchis conjunctus	Canada,[50] Greenland[8]
Metorchis orientalis	China[51]
Opisthorchis felineus	The Baltic States,[8] eastern Germany,[8] Italy,[52] Kazakhstan,[8] Poland,[8] Russia,[8] Eastern Siberia,[8] Ukraine[8]
Opisthorchis noverca	India[53]
Opisthorchis viverrini	Cambodia,[8] Lao PDR,[8] Thailand,[8] Central and South Vietnam[54]
Troglotrematidae	
Nanophyetus salmincola	Oregon-USA[55]
Nanophyetus s. schikhobalowi	Russia[56]
[c-39]*Klick, Tantachamroon*	

infection, which is still unconfirmed in those regions (Table 1, Appendix A).[46]

The infection of *Clonorchis sinensis*, the Chinese or oriental liver fluke, has mainly been reported in humans in six countries: China, Japan, Korea, Russia, Taiwan, and North Vietnam. Recently, infection by *Clonorchis sinensis* was detected in fecal samples of nine

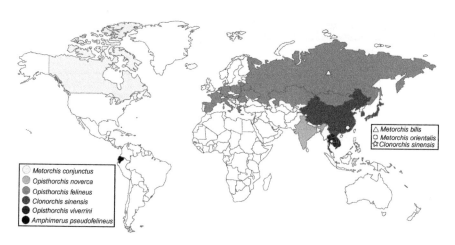

Figure 1.1 Geographical distribution of the liver fluke in the family Opisthorchiidae, comprised of Metorchis, Opisthorchis, Clonorchis, *and* Amphimerus.

individuals from central Thailand by PCR-RFLP.[48] This was the first report of clonorchiasis cases in Thailand, and the only report so far.

Clonorchiasis has affected the Chinese for more than 2000 years. From a 2003 nationwide survey, the infection rate was 2.4% with an estimation of 12.49 million people infected with *C. sinensis* in China. The 2003 survey result also showed that the infection rate of *C. sinensis* increased by 74.8% when compared with the results of a 1990 nationwide survey. The infection rate in males was higher than in females; and the infection rates among people eating raw fish or eating out frequently were higher than those who did not.[59]

Three localities in the central (Sun Moon Lake), middle (Miao-Li), and south (Mei-Nung) of Taiwan are known as heavy endemic areas of clonorchiasis with infection varying from 52–57%.[60] Ultrasonography was carried out in 1,081 people in South Taiwan; 89 of 947 clonorchiasis-infected cases (9.4%) were found to have gallstone formation, with most (85 cases, 8.9%) in the gall bladder. In a non-clonorchiasis group of 144 people; gall stones were detected in 8 patients (5.6%), and stones were found in the gall bladder of 6 cases (4.2%).[61] Rats, cats, dogs, and pigs were found to be the natural reservoir hosts, and the infection rate in pigs was the highest (Table 1.2).[64] However, no recent reports have been published on the current status of clonorchiasis in Taiwan.

Table 1.2 Clinical Manifestations of Small Liver Flukes

Species	Incubation Period	Acute Infection	Chronic Infection
*Clonorchis sinensis	4 weeks	No report	Asymptomatic or with fever, malaise, abdominal pain, rash, gall bladder, gall stones, cholangitis, cholecystitis, liver abscess
Metorchis conjunctus	1–15 days	Low fever, epigastric pain, raised liver enzymes and eosinophil counts	No report
Opisthorchis felineus	2–4 weeks	High fever, increased liver enzymes and eosinophil counts	Cholangitis, liver abscess
Opisthorchis viverrini	4 weeks	No report	Asymptomatic or with hot sensation in abdomen, flatulence, fatigue, cholangitis, jaundice

*The causative agents for Cholangiocarcinoma[8,50,62,63].

Clonorchiasis was considered to be a medically important parasitic problem in 19 prefectures in Japan around 1950. The control programs included a change in land usages, snail control, and health education; these were highly effective, so that in 1991 no *C. sinensis* egg was found in a million fecal sample examinations.[8]

A high prevalence of clonorchiasis was reported in Korea in 1950, with infection rates ranging from 15.9% to 40.2% in 7 river basins in endemic areas.[8] Following the control program, a slight decrease of prevalence was reported recently from the same areas,[65] with the overall infection rate 11.1% with a range of 0.4% to 30.6%. According to the National Cancer Incidence database for 1999–2005, approximately 10% of cholangiocarcinoma in Korea was caused by chronic clonorchiasis: with 12% for males and 6% for females. The incidence increased to a quarter of cases in endemic areas.[66] In 2004, the national survey on intestinal helminthiasis showed an overall 2.9% egg positive rate of *C. sinensis*, with an estimated 1.3 million infected in Korea.[67]

In Russia, it has been estimated that at least 1 million people are infected in the Amur River basin known-endemic area of clonorchiasis.[47]

In Vietnam, *C. sinensis* was found in 9 Northern provinces, with prevalence varying from 0.2% to 26%, and an estimated one million people infected.[54] Summarizing those recent reports on *Clonorchis sinensis* infections, it is globally estimated that more than 200 million people are at risk, 15–20 million people are infected, and 1.5–2 million have symptoms or complications.[68]

The *Metorchis* liver fluke is reported worldwide in fish-eating mammals and birds. The *M. conjunctus* infection has been found in Canada and the northern USA; the reported definitive hosts are carnivores: the raccoon, the fox, the wolf, and the dog. About 1–3% of wolves in Canada have been found to be infected with *M. conjunctus*; cholangiohepatitis with periductular fibrosis in the liver are the lead pathogenesis reported in those wolves.[69] Human infections of *M. conjunctus* were sporadically reported among native peoples in Canada.[70] In 1993, a common-source outbreak of *M. conjunctus* infection was described in 19 people who ate sashimi prepared from white sucker fish caught in a river north of Montreal, Canada. The symptom findings were low fever either continuous or intermittent and epigastric pain, with an incubation period of 1–15 days. The common laboratory findings were raised eosinophil counts and raised liver enzyme concentration. A strong relation between amount of fish eaten and illness was reported.[50]

Recently, by ITS2 based multiplex PCR, specific bands for *Metorchis bilis* (Braun, 1890) were found in stools containing opisthorchid eggs collected from Tomsk, Russia. This primary work showed about 10% *M. bilis* mixed with *Opisthorchis felineus* infections in this region.[49]

Metorchis orientalis (Tanabe, 1921) has been reported in ducks, cats, and dogs in China, Japan, and Korea. Experimental infection showed that adult worms can developed in chickens also. Natural infection in man of *M. orientalis* was first reported in the Ping Yuan County of the Guangdong Province in China; the egg-positive rate in humans was 4.2% (4/95); 12 adult worms were recovered in the stools of 2 patients after they took anthelmintic drugs. The infection rate in animals was high in this endemic area; it was: 66.7% in ducks, 78.6% in cats, 23.5% in dogs, and 87.6% in fish (*Pseudorasbora parva*).[51]

Human infection of *Metorchis* seems underestimated, particularly in eastern Asia and Eurasia where *Clonorchis sinensis* and *Opisthorchis felineus* are predominantly reported, since animal reservoir hosts (the cat and the dog) and a common fish (*Pseudorasbora parva*) in the region are serving as the final host and the second intermediate host for all three species of liver flukes in the area.

Opisthorchis felineus is endemic in Kazakhstan, western Siberia of the Russian Federation, and Ukraine. The infection involves humans and domestic and wild animals. It is estimated that the at-risk population is about 14 million people, and about 1.6 million people are estimated to be infected.[8] Before the Second World War, the infection of *O. felineus* was endemic also in the Baltic States, eastern Germany, and Poland.[8] *O. felineus* was first reported in animals in Italy in 1884, but infection in humans has just been reported in 2004 involving 32 cases in Central Italy.[52] An investigation on opisthorchiasis in the endemic area around 4 lakes—Trasimeno, Boselna, Vico, and Bracciano—showed that among 15 species of fish, only tench (*Tinca tinca*) was positive with metacercaria of *O. felineus* with infection rates from 28% to 95% with over 50% infection of tench found in Bolsena and Bracciana lakes. Among the 5 species of animals examined, eggs of *O. felineus* were found in cats only with infection rates from 26 to 40%.[71]

Opisthorchis noverca has been known to infect dogs and pigs, and occasionally man, in India. In the early 19th century, the taxonomic position of *O. noverca* was confusing; however, later it was discovered that *O. caninus* (Barker, 1911), *Paropisthorchis indicus* (Stephens, 1912), and *Amphimerus noverca* (Braun, 1903) Barker, 1911 are synonyms of *O. noverca*.[53] Reports on this species have not been carried out recently.

Opisthorchis viverrini is endemic in the Mekong Basin, Cambodia; Lao PDR; Thailand; and South Vietnam. A recent report on 6 cases found in Takeo Province in Cambodia[72] led to a survey in 3 villages of Takeo Province involving 1,799 people, and the result showed opisthorchiasis rates in this population were 47.5% (46.4−50.6%).[73] Examination of fish collected from the adjacent area (Lake 500, Kandal Province) showed many species of cyprinids were infected with metacercaria of *O. viverrini*; the species were confirmed by experimental infection and identification of adult worms together with genetic sequences.[74] It seems that the south of Cambodia is an endemic area of *O. viverrini*.

Opisthorchiasis is highly endemic in the lowlands of central and southern Laos, with an overall prevalence of 10.2%.[75] In 1989, an initial survey in the Khong District of southern Laos showed the prevalence of opisthorchiasis to be as high as 80% in some schools. A year after the control program, the prevalence dropped to 5%. No snail control or treatment of the reservoir hosts (dogs and cats) was undertaken.[8] The prevalence of opisthorchiasis has gradually returned to that before the control program (92%).[76]

Opisthorchiasis is highly endemic in northeast Thailand; in 1953, its overall prevalence was 25%, which increased to 35% in 1984. In 1991, it was reported that 7 million people were infected and 45 million people were at risk. Since 1988, a national control program has been implemented and the prevalence of opisthorchiasis had declined to 7% by 2000.[8,77] According to a stool survey using the Kato–Katz method in 2009, small fluke eggs were frequently found in the north and northeast of Thailand. It was estimated that 6 million people were infected with small fluke eggs,[76] which include liver (*O. viverrini*) and/or intestinal flukes (Heterophyids).

Eleven provinces of South Vietnam have been found to be endemic areas of *Opisthorchis viverrini*, high infection rates were in Phu Yen with infection rates ranging from 15.2 to 36.9%.[48]

Infection by small liver flukes is usually asymptomatic or with low or high, intermittent or continuous fever, abdominal pain, and skin rash; but in chronic heavy infection may involve the enlargement of the gall bladder, cholangitis, cholecystitis, and liver abscess. Differences in clinical manifestation have been reported among species of liver flukes (Table 1.2). The association of liver fluke infection and cholangiocarcinoma has been reported in two species: *Clonorchis sinensis* and *Opisthorchis viverrini*.[8,63] The association of the *O. felineus* infection and cholangiocarcinoma has not been reported; moreover, a recent report on the clinical manifestation of *O. felineus* shows it developing as a febrile eosinophilic syndrome with cholestasis rather than a hepatitis-like or rheumatic–like syndrome as previouslyreported.[62,78,79]

1.1.2 Treatment of Liver Flukes

The first choice is praziquantel, a single dose at 40 mg/kg or 25 mg/kg for 3 consecutive days, which gives cure rates of 91% to 100%.[80] The second choice drug is albendazole at 400 mg, given in 2 dividing doses for 7 days, which gives a cure rate at 63%.[81] Recently, tribendimidine has been reported as a promising drug as it gave a cure rate similar to praziquantel during an opisthorchiasis treatment trial.[82]

1.2 OTHER FISH-BORNE TREMATODES

1.2.1 Clinostomatidae

Clinostome metacercaria have been reported in freshwater fish worldwide. A clinostome metacercaria is big: it can be seen with the naked eye. It is either encysted in the muscle of fish with a very thin cyst wall or it moves freely without a cyst wall inside the body cavity of fish. Its body is stout and a whitish-yellow, so it is sometimes called the "yellow grub." After being ingested, the metacercaria easily excysts and attaches to the esophagus or moves to the trachea. It has been known to cause highly irritated infections of the upper digestive/respiratory tract—pharyngitis, laryngitis, or laryngopharyngitis; this disease is commonly known as halzoun or marrara.[3] Human infection with clinostome was first reported in Japan;[4] it has also been reported in Israel,[3] India,[2] and Korea[5,83] (Table 1.1). One case report in Thailand showed that this worm caused irritation to the eye; ophthalmological examination of a man who had a pain in the frontal sinus area for 2−3 months revealed a white spot at the right lower inner eyelid, mild inflammation, and conjunctivitis. After the worm was removed from the lacrimal opening, the patient fully recovered.[6]

1.2.2 Echinostomatidae

Transmission of the echinostome to humans is either through eating raw or undercooked fish, snails, or amphibians. Human cases have been reported mostly in Asia. Echinostomes are morphologically distinct from other trematodes by the presence of spines around the oral sucker, forming a specific shape known as collar spines. The number and arrangement of these spines is the basis of genus identification. Duodenum mucosal bleeding and ulceration are the main clinical findings due to mechanical damages caused by the worms. The common symptoms are abdominal pain and diarrhea followed by weakness and weight loss.[7] The species transmitted through fish are summarized (Table 2, Appendix A), along with the snail, fish, and animal reservoir hosts. The first drug of choice to treat echinostome infections is 10−20 mg kg-1 praziquantel in a single oral dose. Albendazole is the second-line drug.[84]

1.2.3 Heterophyidae

The heterophyid is a small sized fluke, about 1 mm. in length, and is parasitic mostly in the small intestine of birds and mammals and rarely in fish and reptiles. The fluke has a specialized structure at the genital pore, the gonotyl, which forms the basis of the family classification. Infection in man is of great interest since infection outside of the intestine in man has been reported—sometimes in vital organs (i.e., heart, spinal cord, and brain)—by eggs and adult worms embolized in the blood vessels and has been the cause of death.[22] The worms are usually found lodging in intestinal mucosa between villi, however, they have invaded the submucosal level in experimental immunosuppressive mice.[85] Within a week after the metacercaria is ingested by the definitive host, metacercaria develop to mature adults in the intestine.[86] Heterophyid adults have a short life; the reported life spans varied among different host species. In immunosuppressed conditions of the same host strain, worms survived longer.[85] More than 25 species have been found parasitizing humans around the world. The list of species and their distributions and reported hosts are summarized in Table 1.1 and Table 3 of Appendix A. The first line drug is 10–20 mg kg-1 praziquantel in a single oral dose.[84] Albendazole can be considered the second-line drug; a single dose gives around a 48% cure rate.[87]

1.2.4 Troglotrematidae

Nanophyetus salmincola is known as salmon poisoning disease as the fluke's metacercaria is transmitted by salmon. The infection in dogs has been reported as fatal since the fluke acts as a transmitter of rickettsia, *Neoricketsia helminthoeca*, which is the causative agent of virulence disease in dogs.[88] A high prevalence of infection in humans was reported in native peoples of eastern Siberia in 1931 by *Nanophyetus schikhobalowi*.[56] Later the species was changed to a subspecies level since there are morphological similarities between *N. salmincola* and *N. schikhobalowi*.[89]

Between 1974 and 1985, 10 patients were reported to be infected with *N. s. salmincola* in Oregon, USA[90] (Table 4, Appendix A). All of the patients had eaten raw flesh or eggs of the steelhead trout, and/or insufficiently cooked or smoked salmon before the onset of symptoms or positive laboratory findings. The common clinical and laboratory findings were diarrhea and raised eosinophil counts (6%–43%). Two of the patients with positive lab findings had no symptoms. Bithional

(50 mg/kg orally on alternate days x2 doses) and niclosamide (2 g orally on alternate days x3 doses) were found to be effective, but no immediate cure was observed following treatment with mebendazole (100 mg orally twice a day x3 days).[90] At present praziquantel is the drug of choice.

REFERENCES

1. Schell SC. *How to know the trematodes.* Iowa: WMC Brown; 1970.

2. Cameron TWM. Fish-carried parasites in Canada. I. Parasites carried by freshwater fish. *Can J Comp Med* 1945;**9**:245−54.

3. Witenberg G. What is the cause of the parasitic laryngo-pharyngitis in the near East "Halzoun"?. *Acta Med Orient* 1944;**3**:191−2.

4. Yamashita J. *Clinostomum complanatum*, a trematode parasite new to man. *Annot Zool Jpn* 1938;**17**:563−6.

5. Chung DI, Moon CH, Kong HH, et al. The first human case of *Clinostomum complanatum* (Trematoda: Clinostomatidae) infection in Korea. *Korea J Parasitol* 1995;**33**:219−23.

6. Tiewchaloen S, Udomkijdecha S, Suvouttho S, et al. Case report: *Clinostomum* trematode from human eye. *Southeast Asian J Trop Med Public Health* 1999;**30**:382−4.

7. Chai JY. Echinostomes in humans. In: Fried B, Toledo R, editors. *The biology of echinostomes.* New York: Springer; 2009. p. 147−83. Available from: http://dx.doi.org/10.1007/978-0-387-09577-6_7.

8. WHO. *Control of foodborne trematode infections. Reports of a WHO study group.* Geneva: World Health Organization; 1995 (WHO Technical Report Series No 849).

9. Niemi DR, Macy RW. The life cycle and infectivity to man of *Apophallus donicus* (Skrjabin and Lindtrop, 1919) (Trematoda: Heterophyidae) in Oregon. *Proc Helminth Soc Washington* 1974;**41**(2):223−9.

10. Chieffi PP, Gorla MCO, Torres DMAGV, et al. Human infection by *Phagicola* sp. (Trematoda, Heterophyidae) in the municipality of Registro, São Paulo State, Brazil. *J Trop Med Hyg* 1992;**95**:346−8.

11. Tanabe H. Studien uber die Trematoden mit Susswasserfischen als Zwischenwirt. I. Stamnosa armatum n.g., n.sp. *Kyoto Igaku Zasshi* 1922;**19**:1−14.

12. Chai JY, Lee SH. Intestinal trematodes of humans in Korea: *Metagonimus*, heterophyids and echinostomes. *Kisaengch'unghak Chapchi* 1990;(28 Suppl.):103−22.

13. Waikagul J, Wongsaroj T, Radomyos P, et al. Human infection of *Centrocestus caninus* in Thailand. *Southeast Asian J Trop Med Public Health* 1997;**28**:831−5.

14. Looss A. Notizen zur Helminthologie Aegyptens. *Ctbl Bakt* 1896;**21**:913−26.

15. Faust EC, Nishigori M. The life cycles of two new species of Heterophyidae, parasitic in mammals and birds. *J Parasitol* 1926;**13**:91−128.

16. Chai JY, Sohn WM, Yong TS, et al. Centrocestus formosanus (Heterophyidae): human infections and the infection source in Lao PDR. *J Parasitol* 2012. Epub ahead of print.

17. Nishigori M. On a new trematode *Stamnosoma formosanum* n. sp. And its life history. *Taiwan Igakkai Zasshi* 1924;**234**:181−228 [In Japanese].

18. Kobayashi H. Studies on trematodes in Hainan Island. II Trematodes found in the intestinal tracts of dogs by experimental feeding with certain fresh and brackish water fish. *Jpn J med Sci Path* 1942;**6**:187–227.

19. Witenberg G. Trematodiases. In: Van de Hoeden, editor. *Zoonoses*. Amsterdam: Elsevier; 1964. p. 624–6.

20. Khalil M. The life history of a heterophyid parasite in Egypt. *C R Congr Int Med Trop Hyg* 1932;**4**:137–46.

21. Giboda M, Ditrich O, Scholz T, et al. Human *Opisthorchis* and *Haplorchis* infections in Loas. *Trans R Soc Trop Med Hyg* 1991;**85**:538–40.

22. Africa CM, de Leon W, Garcia EY. Visceral complications in intestinal heterophyidiasis of man. *Acta Med Philipp* 1940;1–132 Monographic Series No. 1.

23. Radomyos P, Bunnag D, Harinasuta T. *Haplorchis pumilio* (Looss) infection in man in northeastern Thailand. *Southeast Asian J Trop Med Public Health* 1983;**14**:223–7.

24. Dung DT, van De N, Waikagul J, et al. Fishborne zoonotic intestinal trematodes, Vietnam. *Emerg Infect Dis* 2007;**13**:1828–33.

25. Kuntz RE. Intestinal protozoa and helminthes in school children of Decca, East Pakistan (East Bengal). *Am J Trop Med Hyg* 1960;**8**:561–4.

26. Belizario VY, de Leon WU, Bersabe MJ, et al. A focus of human infection by *Haplorchis taichui* (Trematoda: Heterophyidae) in the southern Philippines. *J Parasitol* 2004;**90**:1165–9.

27. Manning GS, Lertprasert P. Four new trematodes of man from Thailand (letter to editor). *Trans R Soc Trop Med Hyg* 1971;**65**:101–2.

28. Africa CM. Description of three trematodes of the genus *Haplorchis* (Heterophyidae), with notes on two other Philippines members of this genus. *Philipp J Sci* 1938;**66**:299–307.

29. Lie KJ. Some human flukes from Indonesia. *Docum Neerl Indones Morb Trop* 1951;**3**:105–16.

30. Chai JY, Park JH, Han ET, et al. Mixed infections with *Opisthorchis viverrini* and intestine flukes in residents of Vientiane Municipality and Saravane Province in Laos. *J Helminth* 2005;**79**:283–9.

31. Africa CM, Garcia EY. Heterophyid trematodes of man and dog in the Philippines with descriptions of three new species. *Philipp J Sci* 1935;**57**:253–67.

32. Katsuta I. Studies on Formosan trematodes whose intermediate hosts are brackish water fishes. II. *Metagonimus minutus* n.sp., with mullet as its vector. *Taiwan Igakkai Zasshi* 1932;**31**:26–39 [In Japanese, English summary].

33. von Siebold CT. Beitrage zur Helminthographia, humana. *Zeitschr Wiss Zool* 1853;**4**:62–4.

34. Witehberg G. Studies on the trematode family Heterophyidae. *Ann Trop Med Parasitol* 1929;**23**:131–239.

35. Kuntz RE, Lawless DK, Lang Behn HR. Intestinal protozo and helminthes in the people of western (Anatolia) Turkey. *Am J Trop Med Hyg* 1958;**7**:298–301.

36. Ozaki Y, Asada J. A new human trematode, *Heterophyes katsuradai* n.sp. *J Parasitol* 1925;**12**:216–8.

37. Onji Y, Nishio T. On a new species of trematodes belonging to the genus Heterophyes. *Iji Shimbun* 1916;**954**:941–6 [In Japanese].

38. Hsieh HC. Outline of parasitic zoonoses in Taiwan. *Formosan Sci* 1959;**13**:99–109.

39. Saito S, Chai J-Y, Kim K-H, et al. *Metagonimus miyatai* sp. nov. (Digenea:Heterophyidae), a new intestinal trematode transmitted by freshwater fishes in Japan and Korea. *Korean J Parasitol* 1997;**35**:223–32.

40. Yamaguti S. *Synopsis of digenetic trematodes of vertebrate*. Tokyo: Keigaku Publishing; 1971,**1**:620–37, 789–813.

41. Aokage K. Studies on the trematode parasites of brackish water fishes in Chugoku coast of Setonaikai. *Tokyo Iji Shinshi* 1956;**73**:217–24 [In Japanese].

42. Kobayashi H. Recent researches on Japanese fishes which serve as the intermediate host of helminthes. *Proc 5 Pacific Sci Congr (Canada, 1933)* 1934;**5**:4157–63.

43. Alicata JE, Schattenberg OL. A case of intestinal heterophyidiasis of man in Hawaii. *J Am Med Ass* 1938;**110**:1100–1.

44. Kagei N, Oshima T, Ishikawa K, Kihata M. Two cases of human infection with *Stellantchasmus falcatus* Onji & Nishio, 1915 (Heterophyidae) in Kochi Prefecture. *Jpn J Parasitol* 1964;**13**:472–8 [In Japanese].

45. Hasegawa T. Uber die enzystierten Zerkaren in *Pseudorasbora parva*. *Okayama Igakki Zasshi* 1934;**46**:1397–434 [In Japanese, German summary].

46. Calvopina N, Cevallos W, Kumazawa H, et al. High prevalence of human liver infection by *Amphimerus* spp. flukes, Ecuador. *Emerg Inf Dis* 2011;**17**:2331–4.

47. Figurnov VA, Chertov AD, Romanenko NA, et al. Clonorchiasis in the Upper Amur region: biology, epidemiology, clinical presentation. *Med Parazitol (Mosk)* 2002;**4**:20–3.

48. Traub RJ, Macaranas J, Mungthin M, et al. A new PCR-based approach indicates the range of *Clonorchis sinensis* now extends to central Thailand. *PLoS Negl Trop Dis* 2009;**3**:e367. Available from: http://dx.doi.org/10.1371/journal.pntd.0000367.

49. Mordvinov VA, Yurlova NI, Ogorodova LM, et al. *Opisthorchis felineus* and *Metorchis bilis* are the main agents of liver fluke infection of humans in Russia. *Parasitol Int* 2012;**61**:25–31.

50. MacLean JD, Arthur JR, Ward BJ, et al. Common-source outbreak of acute infection due to the North American liver fluke *Metorchis conjuctus*. *Lancet* 1996;**347**:154–8.

51. Lin J, Chen Y, Li Y, et al. The discovery of natural infection of human with *Metorchis orientalis* and the investigation of its focus. *Chin J Zoonoses* 2001;**17**:19–21.

52. Armignacco O, Caterini L, Marucci G, et al. Human illnesses caused by *Opisthorchis felineus* flukes, Italy. *Emerg Inf Dis* 2008;**14**:1902–5.

53. Bisseru B. On the genus *Opisthorchis* Blanchard, 1895, with a note on the occurrence of *O. geminus* (Looss, 1896) in new avian hosts. *J Helminthol* 1957;**31**:187–202.

54. De NV, Murrell KD, Cong LD, et al. The food-borne trematode zoonoses of Vietnam. *Southeast Asian J Trop Med Public Health* 2003;**34**(Suppl. 1):12–34.

55. Eastburn RL, Fritsche TR, Terhune CA. Human intestinal infection with *Nanophyetus salmincola* from salmonid fishes. *Am J Trop Med Hyg* 1987;**36**:586–91.

56. Skrjabin KJ, Podjapolskaja WP. *Nanophyetus schikhobalowi*, n. sp. Ein neuer trematode aus Darm des Menschen. *Zlb Bakt I Orif* 1931;**119**:294–7.

57. Rodriguez JD, Gomez-Lince LF, El Montalvan JA. *Opisthorchis guayaquilensis*. *Rev Ecuat Hig Med Trop* 1949;**6**:11–24.

58. Artigas PT, Perez MD. Consideracoes sobre *Opisthorchis pricei* Foster 1939, *O. guayaquilensis* Rodriguez, Gomez e Montalvan 1949 e *O. pseudofelineus* Ward 1901. Descricao de *Amphimerus pseudofelineus minumus* n. sub. sp.. *Mem Inst Butantan* 1962;**30**:157–66.

59. Chen YD, Zhou CH, Xu LQ. Analysis of the results of 2 nationwide surveys on *clonorchis sinensis* infection in China. *Biomed Environ Sci* 2012;**25**:163–6.

60. Chen CY, Hsieh WC. *Clonorchis sinensis*: epidemiology in Taiwan and clinical experience with Praziquantel. *Arzneimittelforschung* 1984;**34**:1160–2.

61. Hou MF, Ker CG, Sheen PC, et al. The ultrasound survey of gallstone diseases of patients infected with *Clonorchis sinensis* in southern Taiwan. *J Trop Med Hyg* 1989;**92**:108–11.

62. Traverso A, Repetto E, Magnani S, et al. A large outbreak of *Opisthorchis felineus* in Italy suggests that opisthorchiasis develops as a febrile eosinophilic syndrome with cholestasis rather than a hepatitis-like syndrome. *Eur J Clin Microbiol Infect Dis* 2012;**31**:1089–93.

63. Mairiang E, Mairiang P. Clinical manifestation of opisthorchiasis and treatment. *Acta Trop* 2003;**88**:221–7.

64. Wang JS, Tung PC, et al. Studies on the control of zoonotic clonorchiasis. (1). An epidemiological survey in several areas of Taiwan. *Natl Sci Counc Monthly ROC* 1980;**8**:113–22.

65. Cho SH, Lee KY, Lee BC, et al. Prevalence of clonorchiasis in southern endemic areas of Korea in 2006. *Korean J Parasitol* 2008;**46**:133–7.

66. Shin HR, Oh JK, Lim MK, et al. Descriptive epidemiology of cholangiocarcinoma and clonorchiasis in Korea. *J Korean Med Sci* 2010;**25**:1011–6.

67. Kim TS, Cho SH, Huh S, et al. A nationwide survey on the prevalence of intestinal parasitic infections in the Republic of Korea, 2004. *Korean J Parasitol* 2009;**47**:37–47.

68. Hong TS, Fang Y. *Clonorchis sinensis* and clonorchiasis, an update. *Parasitol Int* 2012;**61**:17–24.

69. Wobesor G, Runge W, Stewart RR. *Metorchis conjunctus* (Cobbold, 1860) infection in wolves (Canis lupus), with pancreatic involvement in two animals. *J Wildlife Dis* 1983;**19**:353–6.

70. Watson TG, Freeman RS, Staszak M. Parasites in native people of the Sioux Lookout zone, northwestern Ontario. *Can J Public Health* 1979;**70**:170–82.

71. De Liberato C, Scaramozzino P, Brozzi A, et al. Investigation on *Opisthorchis felineus* occurrence and life cycle in Italy. *Vet Parasitol* 2011;**177**:67–71.

72. Sohn WM, Shin EH, Yong TS, et al. Adult *Opisthorchis viverrini* flukes in humans, Takeo, Cambodia. *Emerg Infect Dis* 2011;**17**:1302–4.

73. Yong TS, Shin EH, Chai JY, et al. High prevalence of *Opisthorchis viverrini* infection in a riparian population in Takeo Province, Cambodia. *Korean J Parasitol* 2012;**50**:173–6.

74. Touch S, Komalamisra C, Radomyos P, et al. Discovery of *Opisthorchis viverrini* metacercariae in freshwater fish in southern Cambodia. *Acta Trop* 2009;**111**:108–13.

75. Rim HJ, Chai JY, Min DY, et al. Prevalence of intestinal parasite infections on a national scale among primary school children in Laos. *Parasitol Res* 2003;**91**:267–72.

76. Sithithaworn P, Andrews RH, De NV, et al. The current status of opisthorchiasis in the Mekong Basin. *Parasitol Int* 2012;**61**:10–6.

77. WHO. *Joint WHO/FAO workshop on Food-borne trematode infections in Asia. Regional office for the Western Pacific.* World Health Organization; 2004 Report Series Number: RS/2002/GE/40(VTN).

78. Belova VG, Paturina NG, Gurevich EN. Early differential diagnosis of icteric variants of viral hepatitis, acute opisthorchiasis and sporadic pseudotuberculosis. *Klin Med* 1981;**59**:73–6.

79. Gordon IB, Razumov VV, Baranova MN, et al. Clinical disguises of opisthorchiasis. *Klin Med* 1984;**62**:108–11.

80. Bunnag D, Harinasuta T. Studies on the chemotherapy of human opisthorchiasis in Thailand. I. Clinical trial of praziquantel. *Southeast Asian J Trop Med Public Health* 1980;**11**:528–30.

81. Pungpak S, Bunnag D, Harainasuta T. Albendazole in the treatment of opisthorchiasis and concomitant intestinal helminthic infections. *Southeast Asian J Trop Med Public Health* 1984;**15**:44−50.

82. Soukhathammavong P, Odermatt P, Sayasone S, et al. Efficacy and safety of mefloquine, artesunate, mefloquine-artesunate, tribendimidine, and praziquantel in patients with Opisthorchis viverrini: a randomised, exploratory, open-label, phase 2 trial. *Lancet Infect Dis* 2011;**11**:110−8.

83. Park CW, Kim JS, Joo HS, et al. A human case of *Clinostomum complanatum* infection in Korea. *Korea J Parasitol* 2009;**47**:401−4.

84. Chai JY. Intestinal flukes. In: Murrell KD, Fried B, editors. *Food-borne parasitic zoonoses: fish and plant-borne parasites, world class parasites*, vol. II. New York: Springer; 2007. p. 53−115.

85. Chai JY, Kim J, Lee SH. Invasion of *Metagonimus yokogawai* into the submucosal layer of the small intestine of immunosuppressed mice. *Korean J Parasitol* 1995;**33**:313−21.

86. Simoes SBE, Barbosa HS, Santos CP. The life cycle of *Ascocotyle (Phagicola) longa* (Digenea: Heterophyidae), a causative agent of fish-borne trematodosis. *Acta Trop* 2010;**113**:226−33.

87. Waikagul J, Watthanakulpanich D, Muennoo C, et al. Efficacy of benzimidazole carbamate on an intestinal fluke co-infected with nematodes. *J Med Assoc Thai* 2005;**88**:233−7.

88. Milleman RE, Knapp SE. Biology of *Nanophyetus salmincola* and "salmon poisoning" disease. *Adv Parasitol* 1970;**8**:1−41.

89. Filimonova LV. Change in the taxonomic rank of *Nanophyetus schikjobalowi* Skrjabin et Podjapolskaja, 1931 (Trematoda: Nanophyetidae). Vses Obsch Gel'mint. *Acad Nauk SSSR, Moscow* 1968;321−9.

90. Eastburn RL, Fritsche TR, Terhune CA. Human intestinal infection with *Nanophyetus salmincola* from salmonid fishes. *Am J Trop Med Hyg* 1987;**36**:586−91.

CHAPTER 2

Collection of Fish-Borne Trematodes from the Final Host

There are three different locations for the habitats of fish-borne trematodes (FBT) in final hosts: the bile duct for family Opisthorchiidae, the small intestine for families Echinostomatidae, Heterophyidae, and Troglotrematidae (*Nanophyetus*), and the upper digestive/respiratory tract for Clinostomatidae.

Collecting adult flukes for morphological study starts with identifying the infection in the individual host. There are four types of techniques used to diagnose infections: 1) Finding FBT eggs in stool, 2) Finding DNA of FBT in stool, 3) Detecting antibody against FBT, and 4) Finding FBT directly in its normal habitat. After finding eggs or DNA of FBT in the stool, treatment follows by giving a purgative to the host; worms in a purged stool are collected and prepared for morphological study. Antibody detection does not ensure that there is a current infection, since it may last a long time after the infection has been eliminated. Diagnosis of infection by egg finding may not always be possible for clinostomatids. Endoscopy should be performed in an individual with persistent chronic cough to look for worms attached to the pharynx/larynx.

2.1 DIAGNOSTIC METHODS

2.1.1 Stool Egg Examination

The common methods used for stool examination are Direct Smear (DS);[1] the Cellophane Thick Smear (CTS),[2] also known as the Kato Method (K) or its modification for quantitative purposes known as the Kato–Katz Method (KK);[3] and the Formalin Ether concentration method (FE).[4] Any method of stool examination can be performed to detect eggs of fish-borne trematodes (FBT) and can be selected according to the objective of the study, including whether it is quantitative or not. However, each method has limitations and the parasite to be studied should be clearly understood. The simple direct smear method is not appropriate for examining light infections of any parasites, in particular species that produce low daily egg outputs, unless the host is

Approaches to Research on the Systematics of Fish-Borne Trematodes. DOI: http://dx.doi.org/10.1016/B978-0-12-407720-1.00002-9

Table 2.1 Daily Egg Output per Worm of Common Fish-Borne Trematodes		
Family	Species	Eggs/worm/day
Echinostomatidae	*Echinostoma hortense*	1,478[6]–3882[7]
Heterophyidae	*Haplorchis taichui* *Metagonimus yokogawai*	82[8] (168, 46, 14) 280[9]
Opisthorchiidae	*Opisthorchis viverrini*	1,000–2,500[10]
	Clonorchis sinensis	4,000[11]

infected with a large number of worms. For instance, *Haplorchis taichui* – a common minute intestinal fluke in the greater Mekong sub-region (GMS) produces 82 (168, 46, 14) eggs per worm per day in light, moderate, and heavy infection groups.[5] Based on the average fecal excrete per day of 140–150 gm, it would take 1,750 (820–10,000) worms to produce eggs for detecting 1 egg per 1 milligram fecal smear of any smears performed by the direct smear (DS) method. The daily egg outputs of common FBTs are summarized in Table 2.1.

The amount of daily egg production of parasites is subject to variation due to maturity, the crowding effect, or the fluctuation in any consecutive duration. Older aged worms produce smaller numbers of eggs per day than younger mature ones. The daily egg output of worms in the light infection group is higher than those of worms in the moderate or heavy infection groups.[12] In quantitative studies—to reduce a false negative in stool examinations—stools should be collected on a few consecutive days to examine for parasite eggs.

The amount of feces required for each method is another key factor of successful performance; it is recommended that a method with a high amount of fecal sample be used, such as a tick smear or concentration method like K, KK, or FE. Small fluke eggs, which includes those that are approximately 25–30 µm in length, can be difficult for an inexperienced microscopist to observe in a thick smear slide.

The type of laboratory setting makes a difference; for examinations taking place in field stations with only basic equipment, with no option for method selection, DS or TS can be used with an understanding of the disadvantages or limitations of the methods. In a fully equipped laboratory, a concentration method should be used to ensure the maximum result.

Eggs of FBT are shown in Fig. 2.1.

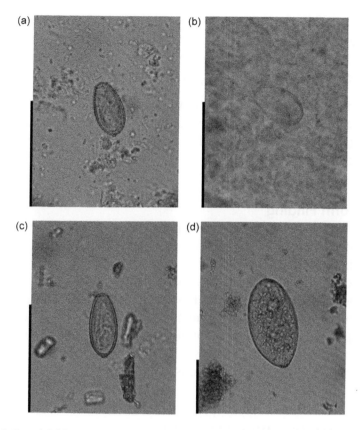

Figure 2.1 Egg of fish-borne trematodes, a. Opisthorchis egg, b. Opisthorchis egg (Kato thick smear), c. Haplorchis egg, d. echinostome egg. Scale bar is 50 mμ.

2.1.2 Copro-DNA Detection

This method is the most sensitive when compared to stool examination; it can detect one egg present in the test sample. However, it is less sensitive than expelled worms in the group that have a small worm load.[13] The method is also highly specific; it has been designed to detect only the desired species of parasite for which DNA sequences are known. Specific primers have been designed to discriminate species of small liver flukes and minute intestinal flukes that coexist in endemic areas.[12,14−20]

2.1.3 Antibody Detection

Several immunological diagnostic methods based on antigen−antibody detection measuring the experience of being infected with FBT have been developed. The specificity of the method depends on the quality of the crude antigens which are partially purified or extracted from adult worms secretion of the worm and the number of heterologous sera used to check cross reactivity with other parasitic infections.[21–23]

This method has a drawback in that it is unable to determine if the diagnosed infection is present currently or in the past, because antibody against the infection may last for a long time after the infection is eliminated.

2.1.4 Worm Finding

In the case of clinostomiasis, infection occurs in the upper digestive/respiratory tracts and is diagnosed by endoscopy, followed by worm removal. It is recommended that if persistent acute pharyngitis or laryngitis occurred after eating uncooked fish, an endoscopy should be performed. Endoscopy is carried out in a patient in a drug-induced sleep, and the surface of the larynx is inspected for worms. Local anesthetic spray of 8% lidocaine solution should be administered if the worm moves rapidly or attaches firmly to the mucosa, to allow the worm to be easily removed by endoscopic forceps.[24]

2.2 TREATMENT AND PURGATION

Praziquantel is highly effective in treatment of FBT, and is considered a first-line drug; the secondline drug is albendazole. Up to this point in time, drug resistance to praziquantel for FBT has not been reported. The recommended dosages are a single dose of 25 mg/kg body weight for intestinal flukes and a single dose of 40 mg/kg or 25 mg/kg, 3 times a day up to 2 consecutive days for liver flukes.[10,25] The worms should not be left to degenerate in a natural bowel movement; a purgative must be used after the treatment.

Praziquantel occasionally causes abdominal discomfort, nausea, headache, dizziness, drowsiness, and tachycardia.[10] Treatment and purgation must be done under close supervision of a physician. The patient should have a light less fibrous meal for dinner; early in the morning without breakfast, the patient should take 40 mg/kg praziquantel, then 2 hours later take 60 ml of a saturated magnesium sulphate solution followed by plenty of warm drinking water (Fig. 2.2).

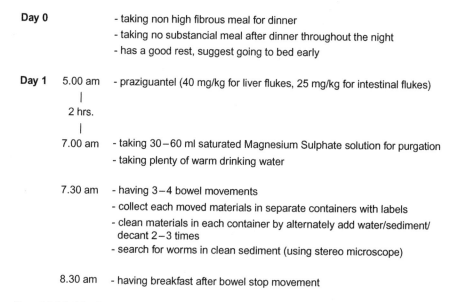

Day 0		- taking non high fibrous meal for dinner
		- taking no substantial meal after dinner throughout the night
		- has a good rest, suggest going to bed early

Day 1 | 5.00 am | - praziguantel (40 mg/kg for liver flukes, 25 mg/kg for intestinal flukes)

|
2 hrs.
|

7.00 am — - taking 30–60 ml saturated Magnesium Sulphate solution for purgation
- taking plenty of warm drinking water

7.30 am — - having 3–4 bowel movements
- collect each moved materials in separate containers with labels
- clean materials in each container by alternately add water/sediment/ decant 2–3 times
- search for worms in clean sediment (using stereo microscope)

8.30 am — - having breakfast after bowel stop movement

Figure 2.2 Schedule of purgative process for worm collection after treatment.

Breakfast should be eaten after the bowel movement stops. Materials from the entire bowel moved are collected and sediment is separated out after repeated washing with clean water for a minimum length of time. Normal saline solution is added to the clean sediment, and an examination for worms under a stereomicroscope is performed or the substance is preserved in 70% alcohol or 5% formalin if the worms cannot be expelled immediately. The expelled worms must be cleaned thoroughly, then preserved with formalin for further staining for morphological study or with alcohol for molecular study. The worm container must be carefully labeled; information about patient, locality, and other information relating to the worm transmission must be fully recorded.

2.3 COLLECTION OF ADULT FBT FROM ANIMALS

The organ habitats of FBT are removed from animal carcasses; each organ is placed on a separate tray. Investigation of liver flukes in the liver is performed by carefully cutting open the bile duct and gall bladder with small pointed-end scissors, alternately washing with normal saline solution (NSS) to avoid damaging the worm. The worms are placed in a small petri dish or a block glass contained NSS.

To complete the collection, place the tray with the cut liver in an incubator at 37°C for 15–30 minutes to encourage the unseen worm to move out from the small bile duct.

Cut the intestines into two or three pieces according to intestinal length and place each part in a separate tray with label (anterior, middle, or posterior). Each species of the recovered worms must have its specific location niche recorded. Investigation for worms in the intestines can be done two ways, 1) Turn the intestine inside out without cutting it open, put it into a bottle containing NSS, close the bottle's cap tightly, and shake vigorously for five minutes or until the intestine is clean from fecal materials. Pour all contents into a beaker, wash and sediment if necessary, and examine the content for worms by stereomicroscope. 2) Cut open along the length of the intestine, wash with NSS, and look for worms under stereomicroscope.

Clean the collected worms by NSS, and fix them in warm 5–10% formalin without pressure; for thick worms slight pressure may be necessary, but the worm must not be flattened too much since internal organs will be disoriented and will not reflect their true locations in the worm body. After that put the solution into a tightly capped bottle with a label (species of worm, location of worm, host species, sex and age of the host, locality of host, fixative, date of collection, collector).

2.4 WORM PREPARATION FOR MORPHOLOGICAL STUDY

Morphological study of FBT is usually carried out on microscropic slides of stained worms. The most popular stain is Semichon's aceto-carmine; the worm should be a little overstained and then decolorized in 1% hydrochloric acid in 70% alcohol. Sometimes counter staining with Fast Green is needed for staining spines on the tegument. A natural mounting medium, such as Canada balsam is recommended for a durable preparation (Fig. 2.3). The appropriate number of worms to study for an accurate measurement is around 20 worms or more.

2.5 DIAGNOSTIC CHARACTERISTICS OF FBT

FBT can be first identified by adult location in the host: If found in the mouth or esophagus of a bird, it is a clinostome; if found in the bile duct of bird or mammal, it is an opisthorchid; and if found in the

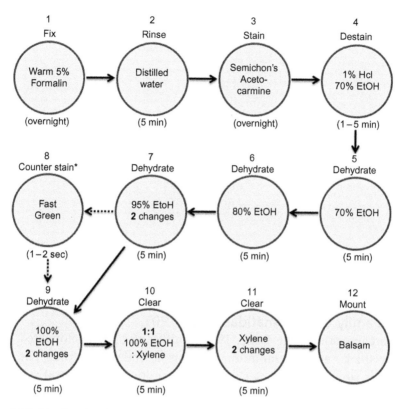

*Figure 2.3 Staining flow chart for making permanent slide of fish-borne trematode. *To stain tegumental and/or oral spines.*

intestines, it is an echinostome, a heterophyid, or a nanophyetid. They can easily be distinguished by family as follows:

1a. Anterior end with collar...2
1b. Anterior end without collar... 3
2a. Collar-like fold..Clinostomatidae
Luhe, 1901
2b. Collar armed with spines... .... Echinostomatidae
Looss, 1902
3a. Genital or ventrogenital sac present...................... Heterophyidae
Leiper, 1909
3b. Ventral simple sucker...4
4a. Genital pore anterior to ventral sucker.............. Opisthorchiidae
Braun, 1901
4b. Genital pore posterior to ventral sucker... ....Troglotrematidae
Odhner, 1914

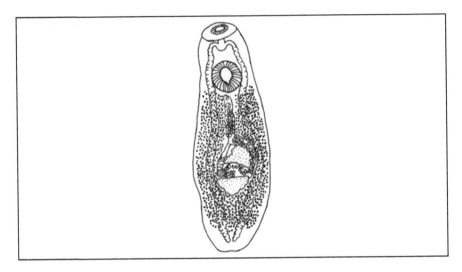

Figure 2.4 Clinostomum complanatum (Rudolphi, 1814) after Caffara et al., 2011.[26]

2.5.1 Family Clinostomatidae Luhe, 1901

In the Clinostomatidae Luhe, 1901 family, the body is elongated, flat, rather thick, and the anterior end has a collar-like fold. The oral sucker is small and the intestinal caeca are long. The ventral sucker is large in the anterior third of the body; the testes are tandem and posterior to the ventral sucker. Cirrus sacs are present; a genital pore is anterior to the anterior testis. There is an ovary between the testes, a short uterus opens into a uterine sac, and vitelline follicles extend from the ventral sucker to the posterior end of body.

Clinostomum complanatum (Rudolphi, 1814), the species reported in man, has an anterior testis located in the posterior end of the middle third of the body, and a posterior testis located in the posterior third of the body. A cirrus sac is in intertesticular space; the uterine sac is tubular; and vitellaria extend from the posterior ventral sucker to the posterior half of the body (Fig. 2.4).

2.5.2 Family Echinostomatidae Looss, 1902

The body in this family is elongated and spinous; the anterior end is provided with collar expansion arms with one or two rows of big spines that interrupt ventrally. The oral sucker is small and the intestinal caeca are long. The ventral sucker is large and in the anterior third of the body; the testes are tandem and posterior to the ventral sucker. A cirrus sac is present; the genital pore is anterior to the ventral

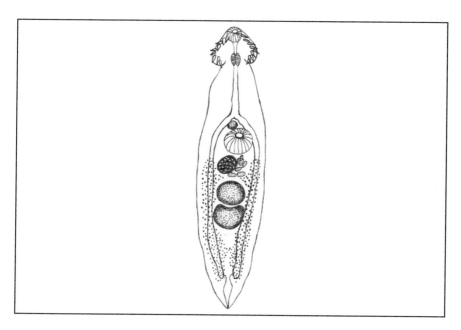

Figure 2.5 Echinochasmus japonicas Dietz, 1909 after Chai, 2009.[27]

sucker; an ovary is anterior to the testes, and a uterus is relatively short and confined to the area between the ovary and the ventral sucker. Vitelline follicles extend from the ventral sucker to the posterior end of the body along caeca.

By the collar spine number and arrangement, the human fish-borne echinostomes can be identified to genera as follows:

1a. Collar spines in one row... 2
1b. Collar spines in two alternative rows, with 4-6 end group...........
                                                         Echinoparyphium Dietz, 1909
2a. Collar spines do not interrupt dorsally, with end group... ....
                                                         Echinostoma Rudolphi, 1809
2b. Collar spines interrupt dorsally, without end group...................
                                                      Echinochasmus Dietz, 1909
                                                                    (Fig. 2.5)

2.5.3 Family Heterophyidae Leiper, 1909

In the Heterophyidae Leiper, 1909 family, the bodies are small and spinous. The oral sucker is simple or modified or circular with rows of spines and caeca short or long. A genital or ventrogenitral sac is present. A ventral sucker is present. A gonotyl is present or absent; a

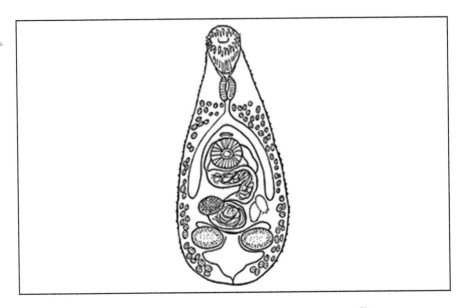

Figure 2.6 Centrocestus Looss, 1899 Centrocestus formosanus after Waikagul et al., 1990.[28]

genital pore is close to the ventral sucker. There are either one or two testes, which are tandem or opposite in the posterior body. The ovary is anterior to the testes, and the uterus is relatively long and posterior to the ventral sucker, with vitelline follicles varying on both lateral sides or dorsal to the posterior body.

Key to the 12 genera of heterophyid flukes reported in humans are:

1a. Anterior end surrounded by 2 rows of circle spines... 2
1b. Anterior end without circle spines... 3
2a. Oral sucker simple... .... *Centrocestus* Looss, 1899
(Fig. 2.6)
2b. Oral sucker with posterior conical appendage... ....*Ascocotyle*
Looss, 1899
(Fig. 2.7)
3a. Ventral sucker sublateral............. *Metagonimus* Katsurada, 1912
(Fig. 2.8)
3b. Ventral sucker median... 4
4a. Seminal vesicle proximal chamber thick-walled and an expulsor... 5
4b. Seminal vesicle chambers thin-walled... 6
5a. Ventral sucker unarmed, gonotyl armed with minute spines, testis single... .... *Procerovum* Onji & Nishio, 1910
(Fig. 2.9)

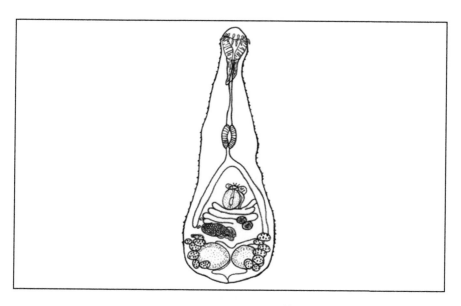

Figure 2.7 Ascocotyle Looss, 1899 Ascocotyle longa after Scholz, 1999.[29]

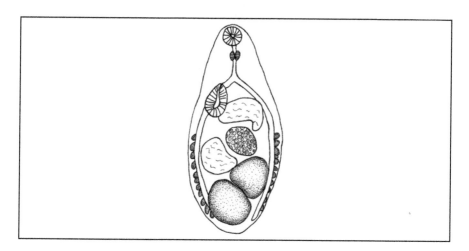

Figure 2.8 Metagonimus, Katsurada, 1912 Schell, 1970.[30]

5b. Ventral sucker small armed with minute spines, unspined gonotyl, testes opposite or single... ....*Stellantchasmus* Onji &
Nishio, 1916
(Fig. 2.10)
6a. One testis... ....*Haplorchis* Looss, 1899
(Fig. 2.11)
6b. Two testes...7

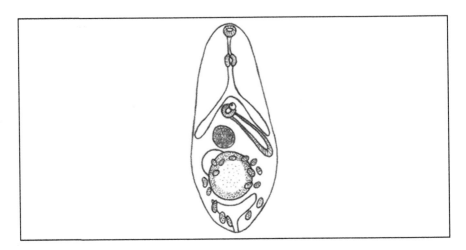

Figure 2.9 Procerovum Onji & Nishio, 1910 after Pearson, 2001.[31]

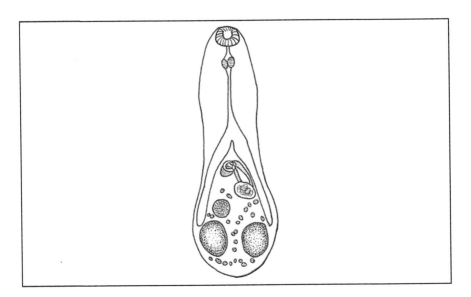

Figure 2.10 Stellantchasmus Onji & Nishio, 1916 after Pearson, 2001.[31]

7a. Gonotyl sucker-like, armed with circle of sclerites, posterosinis-
tral to ventral sucker...8

7b. Gonotyl present or absent...9

8a. Testes opposite................................*Heterophyes* Cobbold, 1899
(Fig. 2.12)

8b. Testes tandem............*Heterophyopsis* Tubangui & Africa, 1938
(Fig. 2.13)

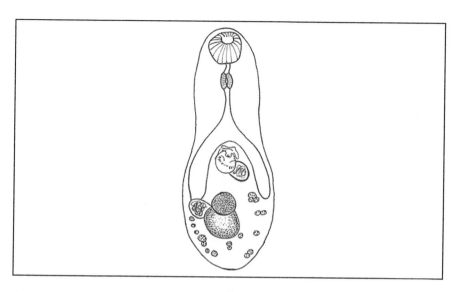

Figure 2.11 Haplorchis *Looss, 1899 after Pearson, 2001.*[31]

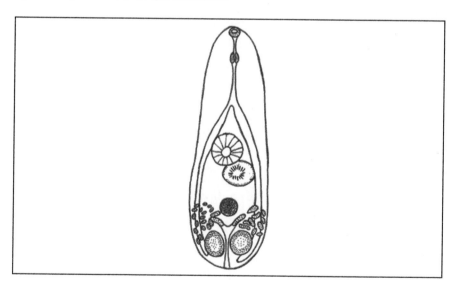

Figure 2.12 Heterophyes *Cobbold, 1899 after Waikagul &Pearson, 1989.*[32]

9a. Gonotyl absent, vitellaria extend anterior to ventral sucker... ....
Cryptocotyle Lühe, 1899
(Fig. 2.14)

9b. Gonotyl present... 10

10a. Gonotyl two knobs, opposite, anterior to ventral sucker... ....
Apophallus Lühe, 1909
(Fig. 2.15)

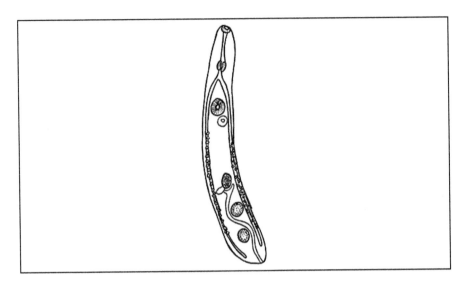

Figure 2.13 Heterophyopsis Tubangui & Africa, 1938 after Pearson, 2001.[31]

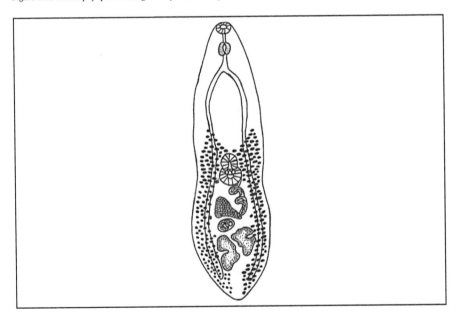

Figure 2.14 Cryptocotyle Lühe, 1899 after Pearson, 2001.[31]

10b. Gonotyl modified, sinistral to ventral sucker..........................11

11a. Ventral simple sucker........................... *Pygidiopsis* Looss, 1907

(Fig. 2.16)

11b. Ventral sucker armed with spines... ....*Stictodora* Looss, 1899

(Fig. 2.17)

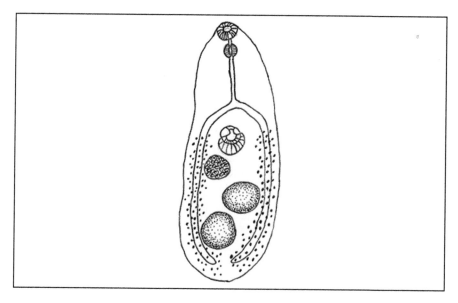

Figure 2.15 Apophallus Lühe, 1909 after Schell, 1970.[30]

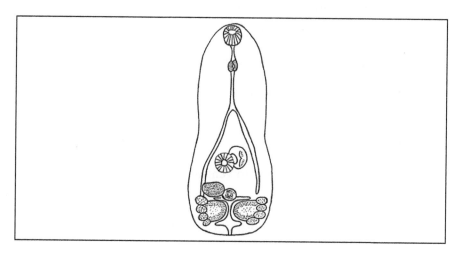

Figure 2.16 Pygidiopsis Looss, 1907 after Pearson, 2001.[31]

2.5.4 Family Opisthorchiidae Braun, 1901

In this family, the body is thin, elongated or oval, and translucent. The oral sucker is small, and the caeca are long. The ventral sucker is small. The testes are tandem or oblique, variable in shape, and located in the posterior of the body. A genital pore is immediately anterior to the ventral sucker. The ovary is small, oval or lobed, and anterior to

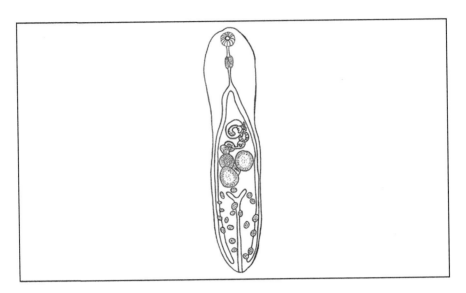

Figure 2.17 Stictodora Looss, 1899 after Pearson, 2001.[31]

the testes. The uterus is long and winding in the mid body between caeca from the ovary to the ventral sucker. Vitelline follicles group or extra caeca are diffuse, in the middle body, or extending to the posterior body. The uterus is omitted from all heterophyid fiqures.

The keys to genera *Opisthorchiid* liver flukes reported in humans:

1a. Testes lobed or oval.. 2
1b. Testes highly branched........................... *Clonorchis* Looss, 1907
(Fig. 2.18)
2a. Vitellaria extend anterior to ventral sucker................. *Metorchis* Looss, 1899
(Fig. 2.19)
2b. Vitellaria in mid body posterior to ventral sucker... 3
3a. Vitellaria not interrupted... .... *Opisthorchis* Blanchard, 1895
(Fig. 2.20)
3b. Vitellaria interrupted at ovary level.... .*Amphimerus* Barker, 1911
(Fig. 2.21)

2.5.5 Family Troglotrematidae Odhner, 1914

In this family, the body is ovoid and spinous. The oral sucker is well developed, and the caeca are long. The ventral sucker is also well developed; the testes are opposite, variable in shape, and located in the posterior body. A genital pore is immediately posterior to the ventral

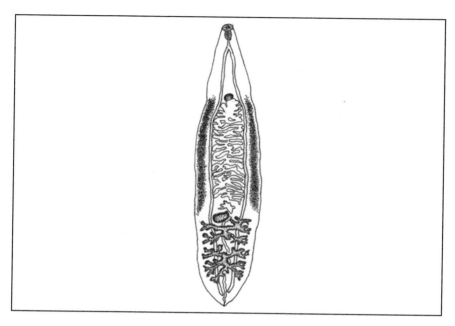

Figure 2.18 Clonorchis Looss, 1907 Scholz, 2001[33].

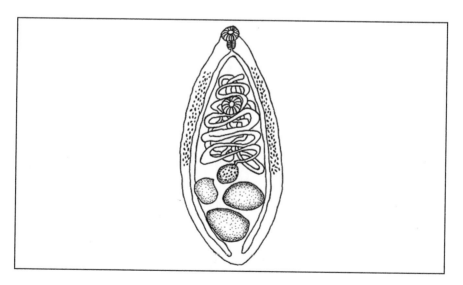

Figure 2.19 Metorchis Looss, 1899 Scholz, 2001.[33]

sucker. An ovary is variable in shape, anterior to the testes, and over-lapping with the ventral sucker. The uterus is short, between the testes and ovary, and vitelline follicles are diffuse in the lateral body from the anterior to the posterior body.

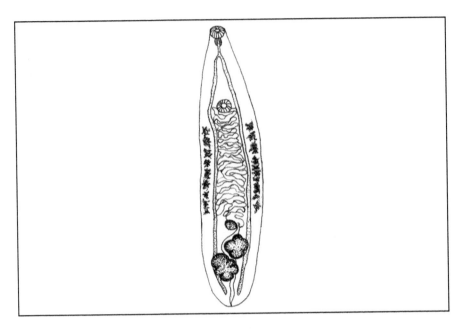

Figure 2.20 Opisthorchis Blanchard, 1895 after Scholz, 2001.[33]

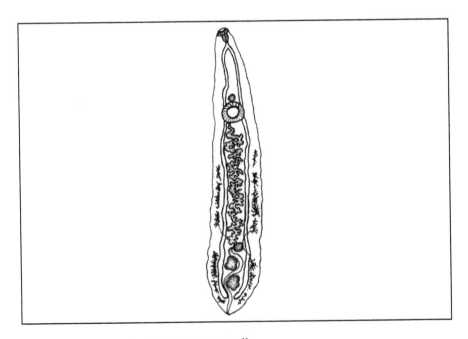

Figure 2.21 Amphimerus Barker, 1911, after Scholz, 2001.[33]

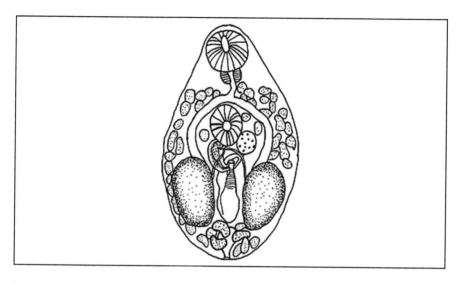

Figure 2.22 Nanophyetus Chapin, 1927 after Blair, Tkach and Barton, 2001.[34]

In this family, there is only one species with two subspecies—*Nanophyetus salmincola* and *N. s. schikhobalowi*—reported in humans. The worm is small and ovoid; the suckers and pharynx are well developed, and the caeca is long. The testes are oval, opposite, and in the posterior body. A cirrus sac is present, dorsodextral to the ventral sucker. The ovary is oval, smaller than the testes, and is dorsosinistral to the ventral sucker. The uterus is short, and is located in between the testes. Vitelline follicles diffuse laterally from the pharynx to the posterior end (Fig. 2.22)

REFERENCES

1. Beaver PC. Quantitative hookworm diagnosis by direct smear. *J Parasitol* 1949;**35**:125–35.

2. Komiya Y, Kobayashi A. Evaluation of Kato's thick smear technique with a cellophane cover for helminth eggs in faeces. *Jpn J Med Sc Biol* 1966;**19**:59–64.

3. Katz N, Chaves A, Pelegrino J. A simple device for quantitative stool thick smear technique in *Schistosomiasis mansoni*. *Rev Inst Med Trop Sao Paulo* 1972;**14**:397–400.

4. Ritchie LS. An ether sedimentation technique for routine stool examination. *Bull Unit Stat Arm Med Depart* 1948;**4**:326.

5. Sato M, Sanguankiat S, Pubampen S, et al. Egg laying capacity of *Haplorchis taichui* (Digenea: Heterophyidae) in humans. *Korean J Parasitol* 2009;**47**:315–8.

6. Lee SK, Chung NS, Ko IH, et al. An epidemiological survey of *Echinostoma hortense* infection in Chongsong-gun, Kyongbuk Province. *Korean J Parasitol* 1988;**26**:199–206.

7. Sohn WY, Huh S, Lee SU, et al. Intestinal trematode infections in the villagers in Koje-myon, Kochang-gun, Kyongsangnam-do, Korea. *Korean J Parasitol* 1994;**32**:149–55.

8. Sato M, Sanguankiat S, Pubampen S, et al. Egg laying capacity of *Haplorchis taichui* (Digenea: Heterophyidae) in humans. *Korean J Parasitol* 2009;**47**:315–8.

9. Yoshida Y. *Medical zoology*. 2nd ed. Tokyo, Japan: Nanzando; 1992, 130–1 (in Japanese).

10. WHO. *Control of foodborne trematode infections. Reports of a WHO study Group*. Geneva: World Health Organization; 1995 (WHO Technical Report Series No 849).

11. Kim JH, Choi MH, Bae YM, et al. Correlation between discharged worms and fecal egg counts in human clonorchiasis. *PLOS-NTD* 2011;**5**:e1339.

12. Sato M, Thaenkham U, Dekumyoy P, et al. Discrimination of *O. viverrini, C. sinensis, H. pumilio*, and *H. taichui* using nuclear DNA-based PCR targeting ribosomal DNA ITS regions. *Acta Tropica* 2009;**109**:81–3.

13. Sato M, Pongvongsa T, Sanguankiat S, et al. Distribution pattern of *Opisthorchis viverrini* and *Haplorchis taichui* infections in southern Laos 2013, Submitted.

14. Le TH, De VN, Blair D, et al. *Clonorchis sinensis* and *Opisthorchis viverrini*: development of a mitochondrial-based multiplex PCR for their identification and discrimination. *Exp Parasitol* 2006;**112**:109–14.

15. Sato M, Pongvongsa T, Sanguankiat S, et al. Copro-DNA diagnosis of *Opisthorchis viverrini* and *Haplorchis taichui* infection in an endemic area of LAO PDR. *Southeast Asian J Trop Med Public Health* 2010;**41**:28–35.

16. Stensvold CR, Saijuntha W, Sithithaworn P, et al. Evaluation of PCR based coprodiagnosis of human opisthorchiasis. *Acta Trop* 2006;**97**:26–30.

17. Thaenkham U, Visetsuk K, Dung DT, et al. Discrimination of *Opisthorchis viverrini* from *Haplorchis taichui* using COI sequence marker. *Acta Trop* 2007;**103**:26–32.

18. Wongratanacheewin S, Pumidonming W, Sermswan RW, et al. Development of a PCR-based method for the detection of *Opisthorchis viverrini* in experimentally infected hamsters. *Parasitol* 2001;**122**:175–80.

19. Wongratanacheewin S, Pumidonming W, Sermswan RW, et al. Detection of *Opisthorchis viverrini* in human stool specimens by PCR. *J Clin Microbiol* 2002;**40**:3879–80.

20. Pauly A, Schuster R, Steuber S. Molecular characterization and differentiation of opisthorch-iid trematodes of the species *Opisthorchis felineus* (Rivolta, 1884) and *Metorchis bilis* (Braun, 1790) using polymerase chain reaction. *Parasitol Res* 2003;**90**:409–14.

21. Sirisinha S, Chawengkirttikul R, Haswell-Elkins MR, et al. Evaluation of a monoclonal antibody-based enzyme-linked immunosorbent assay for the diagnosis of *Opisthorchis viverri-ni* infection in an endemic area. *Am J Trop Med Hyg* 1995;**52**:521–4.

22. Waikagul J, Dekumyoy P, Chaichana K, et al. Serodiagnosis of human opisthorchiasis using cocktail and electroeluted Bithynia snail antigens. *Parasitol Int* 2002;**51**:237–47.

23. Nockler K, Dell K, Schuster R, Voigt WP. Indirect ELISA for the detection of antibodies against *Opisthorchis felineus* (Rivolta, 1884) and *Metorchis bilis* (Braun, 1970) in foxes. *Vet Parasitol* 2003;**110**:207–15.

24. Park CW, Kim JS, Joo HS, et al. A human case of *Clinostomum complanatum* infection in Korea. *Korean J Parasitol* 2009;**47**:401–4.

25. WHO. Report of the Joint WHO/FAO Workshop on Food-borne Trematode Infections in Asia. Ha Noi, Vietnam, 26–28 November 2002. (WHO Technical Report Series No RS/2002/GE/40 ZVTN).

26. Caffara M, Locke SA, Gustinelli A, et al. Morphological and molecular differentiation of *Clinostomum complanatum* and *Clinostomum marginatum* (Digenea: Clinostomidae) metacercariae and adults. *J Parasitol* 2011;**97**:884–91.

27. Chai JY. Echinostomes in humans. In: Fried B, Toledo R, editors. *The biology of echinostomes.* New York: Springer; 2009. p. 147–83.

28. Waikagul J, Vidiassuk K, Sanguankait S. Study on the life cycle of *Centrocestus caninus* (Leiper, 1913) (Digenea: Heterophyidae) in Thailand. *J Trop Med Parasitol* 1990;**13**:50–6.

29. Scholz T. Taxonomis study of *Ascocotyle (Phagicola) longa* Ransom, 1920 (Digenea: Heterophyidae) and related taxa. *Syst Parasitol* 1999;**43**:147–58.

30. Schell SC. *How to know the trematodes.* Iowa: WMC Brown; 1970.

31. Pearson J. Family heterophyidae leiper, 1909. In: Bray RA, Gibson DI, Jones A, editors. *Keys to the trematoda,* vol. 3. London: CABI and Natural History Museum; 2001. p. 113–41.

32. Waikagul J, Pearson JC. *Heterophyes nocens* Onji and Nishio, 1916 (Digenea: Heterophyidae) from the water rat, *Hydromys chrysogaster* Geoff, 1804 in Australia. *Syst Parasitol* 1989;**13**:53–61.

33. Scholz T. Family opisthorchiidae looss, 1899. In: Bray RA, Gibson DI, Jones A, editors. *Keys to the trematoda,* vol. 3. London: CABI and Natural History Museum; 2001. p. 9–49.

34. Blair D, Tkack VV, Barton DP. Family troglotrematidae odhner, 1914. In: Bray RA, Gibson DI, Jones A, editors. *Keys to the trematoda,* vol. 3. London: CABI and Natural History Museum; 2001. p. 277–89.

CHAPTER 3

Collection of Fish-Borne Trematode Cercaria

The cercaria is a free living stage in the life cycle of the trematode between the first intermediate host and the second intermediate host stages. During development, it lives in the snail, which is the first intermediate host, and then it emerges from the snail when fully mature. To study cercariae, one must start by/form collecting from the snail host; both terrestrial and aquatic snails can be hosts of trematodes.

The first intermediate host of FBT is an aquatic gastropod, which has a rather high host specificity for any family of trematode. Snail hosts of most FBTs live in freshwater, but some species of family Heterophyidae use the brackish water snail as their host. Each type of snail has a specific ecology niche, which must be understood before the collection begins. The snails can be collected by simply picking by hand freely, scooping, or dredging, depending on the type of study area and the abundance of snails. To estimate the potential transmission of trematodiasis in the study area, snail density and cercaria infection rate, along with the population of the second intermediate host, should be accurately measured and calculated. Only the collection of snails for cercarial morphological study is described here.

3.1 SNAIL HOST OF FISH-BORNE TREMATODES

Reported snail hosts of echinostome, heterophyid, and opisthorchid trematodes that infect humans are listed in Appendix A.

3.1.1 Family Bithyniidae

The bithyniid snail host of the opisthorchid fluke is small: about 6–15 mm in length. The shell is ovate-conoidal, brownish to olive-color, and has delicate spiral lines. The operculum is calcareous with a paucispiral nucleus. Bithyniids inhabit rice fields, canals, or shallow water reservoirs. They live on mud or sand beds of slow running clear water or on root of aquatic weeds.[1]

Approaches to Research on the Systematics of Fish-Borne Trematodes. DOI: http://dx.doi.org/10.1016/B978-0-12-407720-1.00003-0

3.1.2 Family Planorbidae

The first reported species of snails that are intermediate hosts to echinostomes are pulmonate snails in the family Planorbidae. They are hermaphroditic and is not operculated. The shell is discoidal, dextral, or sinistral. Pulmonary and genital apertures are situated on the right or left side, and eyes are located at the bases of long, filiform tentacles, which are reddish in color. They inhabit clear freshwater either stagnant or slow running, live on aquatic vegetation, and feed on minute algae and fine organic deposits.[1] The reported first intermediate host of *Clinostoma* was *Helisoma*: ram's horn or a planar snail of the family Planorbidae.[2]

3.1.3 Family Thiaridae

A common snail host of the heterophyid fluke is in the family Thiaridae, which can be found worldwide in fresh and brackish wasters. The shell is elongated conic, turreted, or ovate-conoidal, and is solid with spiral ridges and/or axial ribs and microsculpture. The operculum is corneous, either paucispiral or multispiral. The parthenogenetic female is ovoviviparous; the egg develops to a young snail in a brood chamber and is then released through the birth pore located underneath the right tentacle. Thiarid snails inhabit shallow lakes or canals with slow to immediate flow running water. They bury themselves in mud or sandy bottoms or attach to rock.[1]

3.1.4 Family Semisulcospiridae

Nanophyetus salmincola used *Oxytrema silicula* (Gould) of the family Semisulcospiridae as its first intermediate host. Snails of the family Semisulcospiridae are similar to the family Thiaridae: they are elongated conic in shape and can be found in rivers and streams.[3]

Snails in their habitats are shown in Fig. 3.1. After collection, snails must be cleaned and identified before the process for harvesting the emerged cercariae can be completed.

3.2 CERCARIA MORPHOLOGICAL STUDY

A cercaria develops inside a snail, which is its first intermediate host, then it emerges and has a short free swimming life in the water. To prepare for morphological study, place snails in small dishes

Figure 3.1 Snail hosts of fish-borne trematodes in their natural habitats: a. A Bithynia *snail on the mud bottom of an irrigation canal along a rice field; b.* Melanoides *snails on the mud bottom of a pond; c. A planorbid snail attaches to a vegetable floating in a pond.*

containing clean water in one (Fig. 3.2) or more for three hours under illuminated light, and then observe water for cercaria under a stereomicroscope (Fig. 3.3) or crush the snail and examine for cercaria.

Cercariae of closely related species are similar in body type, and they are difficult to distinguish by morphology only. Several techniques are employed to differentiate the cercariae: using a vital stain to observe the pattern and number of internal gland cells, using silver nitrate to stain papillae on the tegument of the body and the tail of cercaria (chaetotaxy), or using Polymerase Chain Reaction (PCR) and

Figure 3.2 Snails singly placed in water for emergence of cercaria.

Figure 3.3 Cercariae (arrow head) newly emerged from snail.

sequencing to study gene sequences. PCR is the best method for identifying cercariae, if matching adult sequences can be located.

3.3 MORPHOLOGY OF FRESH CERCARIA

Drop water contain a few living cercariae on a slide, and apply a cover slip. If cercaria move too vigorously, touch a blot paper to the rim of the cover slip to adjust the amount of water and then flatten to a relaxed position to allow full observation of the cecariae. During observation, free-hand sketch morphology. Measurement is done in a specimen fixed by hot 5% formalin.

Cercaria comprises two distinct parts: the body and the tail. The body of the cercaria is a miniature of the adult, and the internal organs, which resemble those of the adult, are the suckers, the alimentary tract, and the excretory system. The excretory system contains an excretory bladder located at the posterior end of the body. On each lateral side of the anterior wall, the bladder connects with a collecting tube that runs forward and divides into anterior and posterior tubules. Those tubules further branch into small tubules that end with flame cells. The total number of flame cells of both lateral tubes and anterior and posterior tubules is written in a formula as $2[(n + n + n) + (n + n + n)]$. The formula varies by type/species/family of trematodes. A common form is $2[(3 + 3 + 3) + (3 + 3 + 3)] = 36$.

Specialized organs of the cercaria are gland cells (penetration glands, cystogenous glands, mucoid glands), and a stylet or mouth spear. A Pleurolophocerca cercaria has two or three rows of small spines at its mouth opening. Penetration glands and cystogenous glands are visible in fresh cercaria, without the necessity of staining. Penetration glands appear in pairs, one on each side of the midline of the body; each gland connects to a duct that opens anterior to the oral sucker. Cystogenous glands scatter under the body surface, and secrete a substance to form a cyst wall during the encystation of the cercaria. The mucoid gland is invisible without staining. The tail is a swimming aid; it can be simple or can possess finfolds or forks at the end.

3.4 VITAL STAIN

When observing live cercaria, a 0.1% aqueous solution of neutral red or Nile blue sulphate is used. Fifteen minutes after a drop of this solution is added into 50 ml of water with a cercaria. The excretory system, including frame cells, is easier to observe in stained, rather than unstained specimens. Also, the stain can show the presence of different sizes of granules in penetration glands. Toluidine blue or thyonine can be used to look for the presence of mucoid glands, which are normally invisible.

3.5 CHAETOTAXY

Drop cercariae into a hot 0.5–3% aqueous solution of silver nitrate. Wash in a few changes of distilled water, then expose them to strong light for 1–2 minustes or 15 minutes to direct sun light, and then wash in distilled water. After that process, make a permanent slide by dehydrating, clearing in glycerine and mounting in glycerine jelly or use glycerine and mount in glycerine jelly for a semipermanent slide preparation.[3,4] Papillae on the tegument of a cercaria are clearly visible as brownish dots; the number and arrangement of papillae should be studied systematically at three separate areas: the cephalic zone or area, the body, and the tail.[5] The number and pattern of papillae are regular, except in the area of the lateral body, which is variable and difficult to study (Fig. 3.4).

3.6 POLYMERASE CHAIN REACTION

The morphology of the body and tail determines the cercarial type, and it can also be used to identify the superfamily or family of trematodes. The cercaria of human FBT occur in four morphologic types (Fig. 3.5); each type represents a family of FBT(a-Family Troglotrematidae,

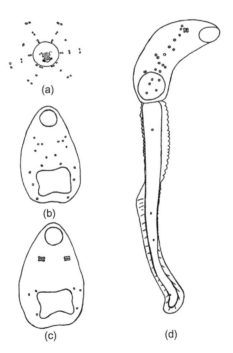

Figure 3.4 Chaetotaxy of Stellantchasmus cercaria (original) showing papillae stained with silver nitrate: (a) Enface view of cercarial mouth, (b) Ventral view of cercarial body, (c) Dorsal view of cercarial body, (d) Lateral view of cercarial body and tail.

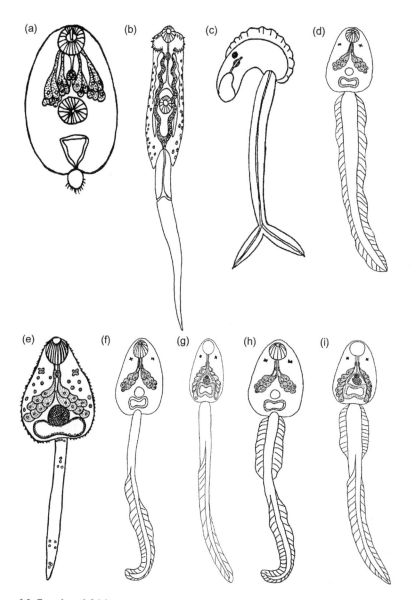

Figure 3.5 **Cercariae of fish-borne trematodes.** *(a) Chaetomicrocercous cercaria (Nanophyetus), after Schell, 1970.[3] (b) Brevifurcate-pharyngeate cercaria (Clinostomum), after Schell, 1970.[3] (c) Echinostome cercaria, after Schell, 1970.[3] (d) Pygidiopsis cercaria, after Ochi, 1931.[6] (e) Centrocestus cercaria, after Waikagul et al., 1990.[7] (f) Pleurolophocercous prevesicular glands, Opisthorchis cercaria (original). (g) Pleurolophocercous paravesicular glands, Stellantchasmus cercaria (original). (h) Parapleurolophocercous prevesicular glands, Stictodora cercaria (original). (i) Parapleurolophocercous paravesicular glands, Haplorchis cercaria (original).*

b-Family Echinostomatidae, c-Family Clinostomatidae), except that the cercariae of Heterophyidae and Opisthorchiidae have closely similar morphologic types (d to i).

Key to FBT cercariae identification (Fig. 3.5)

1a. Tail short, stylet present ....Chaetomicrocercous cercaria
(Family Troglotrematidae) (Fig. 3.5a)

1b. Tail long, stylet absent 2

2a. Anterior end possesses collar spines.........Echinostome cercaria (Family Echinostomatidae) (Fig. 3.5b)

2b. Anterior end without collar spines 3

3a. Long tail with short furca at posterior end; pharynx present; eyespots present; body possesses a dorso-median finfold.........
Brevifurcate-pharyngeate cercaria (Family Clinostomatidae) (Fig. 3.5c)

3b. Tail not bifurcated 4

4a. Tail possesses lateral finfolds, dorso-ventral fin folds present or absent ....Parapleurophocercous cercaria (Family Heterophyidae). 5

4b. Tail without lateral finfolds, dorso-ventral finfolds present ....Pleurophocercous cercaria (Families Opisthorchiidae, Heterophyidae) 6

5a. Lateral finfolds cover the whole length of tail ....
Ascocotyle Looss, 1899; *Pygidiopsis* Looss, 1907   (Fig. 3.5d)

5b. Lateral finfolds cover only anterior portion of tail 7

6a. Penetration glands prevesicular.........*Apophallus* Luhe, 1909; *Centrocestus*[a] Looss, 1899 (Fig. 3.5e); *Clonorchis* Looss, 1907; *Cryptocotyle* Luhe, 1899; *Heterophyes*[b] Cobbold, 1886; *Metagonimus* Katsurada, 1912; *Opisthorchis* Blanchard, 1895   (Fig. 3.5f)

6b. Penetration glands paravesicular......... *Stellantchasmas* Onji & NIshio, 1916   (Fig. 3.5g)

7a. Penetration glands prevesicular......... *Stictodora* Looss, 1899   (Fig. 3.5h)

7b. Penetration glands paravesicular ....*Haplorchis* Looss, 1899; *Procerovum* Onji & Nishio, 1916   (Fig. 3.5i)

---

[a]*Centrocestus formosanus* cercaria has weakly developed dorsoventral finfolds,[8] but the *C. armatus* tail is simple without finfolds.[9]

[b]*Heterophyes aequalis* cercaria has dorsoventral finfolds only,[10] but *H. nocens* cercaria possesses lateral finfolds the whole length of its tail.[11]

REFERENCES

1. Anonymous (TropMed Technical Group). Snails of medical importance in Southeast Asia. *Southeast Asian J Trop Med Public Health* 1986;**17** [282–22].

2. Baker FC. *The molluscan family Planorbidae.* Urbana: University of Illinois Press; 1945.

3. Schell SC. *How to know the trematodes.* Iowa: WMC Brown; 1970.

4. Ginetsinskaya TA, Dobrovolski AA. A new method for finding sensillae in trematode larvae and the significance of the structure in classification. *Dokl Akad Nauk SSSR* 1963;**151**: 460–3.

5. Richard J. La chetotaxie des cercaires. Valeur systematique et phyletique. *Mem Mus Nat Hist Nat Ser A Zool* 1971;**76**:179.

6. Ochi S. Studies on the trematode in brackish water fishes. On the life cycle of *Pygidiopsis summus. Tokyo Lji Shinshi* 1931;**2712**:346–53 [in Japanese].

7. Waikagul J, Vidiassuk K, Sanguankait S. Study on the life cycle of *Centrocestus caninus* (Leiper, 1913) (Digenea: Heterophyidae) in Thailand. *J Trop Med Parasitol* 1990;**13**:50–6.

8. Martin WE. The life histories of some Hawaiian heterophyid trematodes. *J Parasitol* 1958;**44**:305–23.

9. Tanabe H. Studien uber die Trematoden mit Susswasserfischen als Zwischenwirt. I. Stamnosoma armatus n.g., n.sp. *Kyoto Igaku Zasshi* 1922;**19**:1–14.

10. Kuntz RE, Chandler AC. Studies on Egyptian trematodes with special reference to the heterophyids of mammals. II. Embryonic development of *Heterophyes aqualis* Looss. *J Parasitol* 1956;**42**:613–25.

11. Asada J. Studies on the genus Heterophyes prevalent among the Japanese people. Experimental investigations on the life history of *Heterophyes heterophyes. Jikken Igaku Zasshi* 1928;**12** [897–48. (in Japanese)].

CHAPTER 4

Collection of Fish-Borne Trematodes in Infective Stage from the Fish: The Second Intermediate Host

The infective stage of the fish-borne trematode is the encysted form of metacercariae in fish. The researcher/investigator who looks for a particular species of metacercariae must know the fish species of the 2nd intermediate host plus the ecological type of the snail of the 1st intermediate host. For example, if looking for metacercariae of the small liver fluke, the researcher should look in cyprinid fish collected from a rice field, pond, or irrigation canal in the endemic area. Fish from a big river—even in the same district—may be infected, but at a lower concentration. Reported hosts of fish-borne trematodes that infect humans are summarized in Appendix A, Tables 4 and 5.

4.1 DEVELOPMENT OF METACERCARIAE

Upon reaching the fish, a cercaria will penetrate a fish by attaching its body to the fish and discarding its tail. Secretion from penetration glands with the aid of a stylet or spines at the anterior end of a cercaria enable it to invade the fish body and to move to the site where encystation will take place. After the cercaria reaches the site of encystation in the fish, cystogenous glands secrete substances to form a cyst wall to cover the entire cercarial body.

During the initial stage after encystation, vacuoles can be found in the metacercarial body. The vacuole period differs among species of parasite from a few hours to a few days. These vacuoles later disappear and the metacercariae become mature within 3 to 4 weeks. Cysts can be round or oval, depending on the species of the parasite. An encysted metacercaria increases its size gradually during development. Cysts of the *Clonorchis* metacercariae increase in size up to 1.4 times when mature[1] and decreases in size once they are older. A metacercaria can be classified at different levels: by family, as in the opisthorchid liver

Approaches to Research on the Systematics of Fish-Borne Trematodes. DOI: http://dx.doi.org/10.1016/B978-0-12-407720-1.00004-2

fluke, or by genus/species, as in the clinostome, echinostome, or heterophyid.[1]

4.2 METHODS OF FISH EXAMINATION FOR METACERCARIAE

In many families of trematodes, there are both freshwater and brackish water species of fish that act as second intermediate hosts. Fish species must be identified before the fish can be examined for metacercariae. The researcher should learn how to use identification keys to identify local fish. Then they will only need to be sent to an expert if they are an unknown species or for a second opinion on uncertain specimens. Before examining the fish, the researcher must identify, measure, or weigh it in order to analyze the infection rate and density of metacercariae in particular species of fish.

4.2.1 Compression Method
Press the entire fish (for small fish) or a portion of the fish (for big fish) between two glass plates (approx. 4×6 inches) and search for metacercariae under a stereomicroscope. Dissect metacercariae from the fish with needles and then identify them under a light microscope.

4.2.2 Digestion Method
Grind the entire fish with a mortar or a homogenizer, transfer it to a beaker containing artificial gastric juice, put the beaker in an incubator at 37°C for 2 hours (stirring occasionally). Then strain it through a 1×1 mm mesh filter to remove big particles, clean the sediment with normal saline several times, and observe metacercariae in the sediment under a stereomicroscope.

To record the location of the metacercariae in the fish body, the compression method is the more useful technique, but to harvest metacercariae in bulk, select the digestion method. Because it is preferable to use fresh fish, setting up a laboratory for examination of fish at the field site of fish collection is recommended and can easily be done.

4.3 MOBILE LABORATORY FOR FISH EXAMINATION

When working in the field, the compression method is more convenient. The minimum equipment needs for fish examination are a microscope, a stereomicroscope, trays and a cutting board, a knife, dissecting needles, a pipette and rubber bulb, glass plates, block glasses, microscopic slides and cover slips, tight cap small vials, plastic bags, label tape and a pen, a portable ice flask and box, normal saline solution, 70% alcohol, and 10% formalin. A blender may be needed to extract a high number of metacercariae from fish. Other important items are: a digital camera and a picture key of fish and metacercariae (Fig. 4.1).

Steps of examination (Fig. 4.2):

1. Take a photograph of fish and identify fish with key.
2. Measure and record the number of fish in each species, the date, and the locality of the fish examined.
3. Examine the pectoral fin, tail fin, body muscle, and internal organs of the fish with a stereomicroscope.
4. Separate the metacercaria from the fish using dissecting needles, and pipette it into a block glass.
5. Identify the metacercaria by stereomicroscope if possible; if that fails, examine it under a light microscope.
6. Keep the metacercariae in vials with normal saline solution for further study, with 70% alcohol for molecular study, and in 5% formalin for staining.
7. Label with the date, the species of metacercariae and fish (coding if not known), the locality where the fish was found, and the formula of the solution in the vial.

4.4 METHOD OF EXCYSTING METACERCARIAE

The encysted metacercaria can be identified by the shape and size of the cyst and how the metacercarial body folds in the cyst. But to see more detail, the entire body can be seen clearly once the cyst wall has been removed. Details include: flame cells, frontal glands, pharyngeal glands, tegumental glands, tegumental spines, collar spines, and spines on the ventral sucker or gonotyl.

To remove the cyst wall:

1. Physically pierce the cyst wall by hand using sharp dissecting needles. (Fig. 4.3).
2. Alternatively, a researcher can expose the metacercarial cyst to artificial gastric juice for 30 minutes and to artificial intestinal juice for 30 minutes at 37°C.[1] (Fig. 4.4).

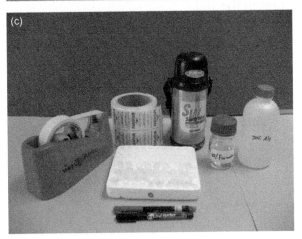

Figure 4.1 Equipment for field laboratory for examining fish
a. *For compression method*
b. *For mincing fish to separate metacerariae*
c. *For keeping metacercariae*

(a)

(b)

(c)

Figure 4.2 (Continued)

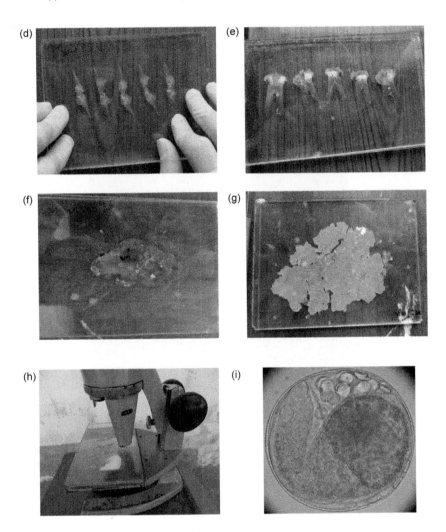

Figure 4.2 Steps of fish examination in the field
a−c. Identify and measure fish
d−h. Examine parts of fish under stereomicroscope
i. Haplorchis taichui: common metacercariae found in Laos PDR and Thailand

4.5 MORPHOLOGY OF METACERCARIAE

4.5.1 Clinostomatidae

Metacercariae of the Clinostome family occures in both encysted and non-encysted forms. The encysted metacercariae are found in the fish muscle; the non-encysted (naked) metacercariae are found in the body cavity, crawling among the internal organs of the fish.

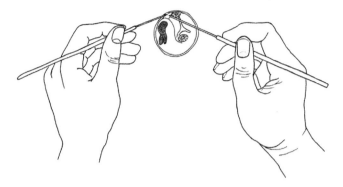

Figure 4.3 Excysting metacercariae with dissecting needles

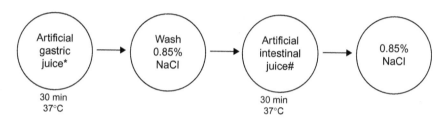

Figure 4.4 Excysting metacercariae with artificial enzymes

The metacercarial body is yellowish, and this type is known as the yellow grub (Fig. 4.5).[2]

4.5.2 Echinostomatidae

Metacercariae of the echinostome family is encysted in snails, fishs, and amphibians. The cyst is oval or round; the metacercariae can be folded or not, and they possess collar spines and corpuscles in their excretory tubes, which is characteristic of echinosomes' cercariae and metacercariae (Fig. 4.6).

4.5.3 Heterophyidae

Heterophyid metacercariae are usually found encysted in many families of fish, such as: Cobitidae, Cottidae, Cyprinidae, Gobiidae, Mugilidae, Ophicephalidae, Percidae, Plecoglossidae, or Siluridae. *Centrocestus formosanus* has been encysted experimentally in frogs.[3]

The heterophyis cyst is usually small and ovoidal. The cyst wall is composed of a thin layer of hyaline material secreted from cystogenous

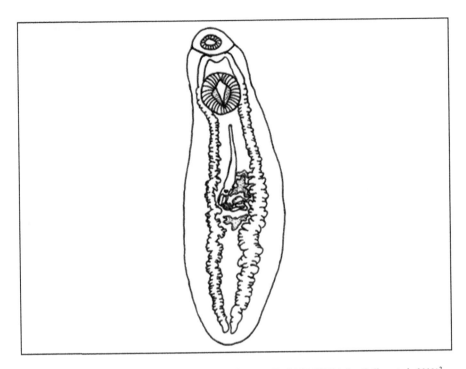

Figure 4.5 Excysted metacercariae of Clinostomum complanatum (Rudolphi, 1814) (after Caffara et al., 2011)[2]

glands in the cercarial body, and it is covered with a thick layer of reaction tissues of the host fish. The fully grown metacercariae look like young nonovigerous adults (Fig. 4.7). Eggs have been reported in the metacercariae of *Stictodora tridactyla*[4] and *Apophallus imperator*.[5]

The majority of heterophyid metacercariae are found encysted throughout the body muscle, fins, and adipose tissue around the internal organs. The site of encystment of some species is very specific, such as that of *Ascocotye diminuta*[6] or *Centrocestus formosanus*,[3,7] where metacercariae are encysted in the gills of the fish. It seems that species in which cercariae have well-developed dorsoventral finfolds, have less-specific sites of metacercarial encystment, whereas species in which cercariae have simple or weakly developed dorsoventral finfolds, have specific sites of metacercarial encystment.

4.5.4 Opisthorchiidae
Metacercariae of all opisthorchiid species have similar morphology. The cyst is oval, there are metacercarial body folds, and the metacercariae move actively. They have oral and ventral suckers, the

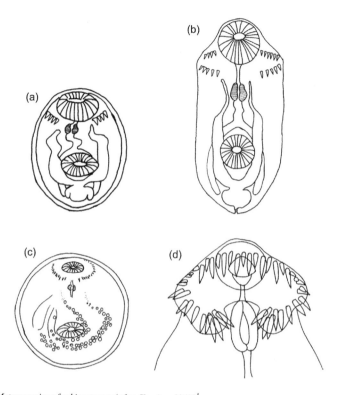

Figure 4.6 Metacercariae of echinostomes (after Komiya, 1965)[1]
a.   *Encysted metacercariae of Echinochasmus japonicus*
b.   *Excysted metacercariae of E. japonicus*
c.   *Encysted metacercariae of Echinoparyphium recurvatum*
d.   *Anterior end of metacercariae of E. recurvatum*

caeca and excretory bladder can clearly be seen; and other organs have not yet developed (Fig. 4.8).[8] The small liver fluke is important; it has been determined that it is a carcinogen that causes cholangiocarcinoma. Many aspects of the biology of metacercariae have been studied.

4.5.4.1 Life Span of Metacercariae

In experiments, mature *Opisthorchis viverrini* metacercariae have been found during 3–27 weeks post cercarial infection with recovery rates of 2.2 (0.1–4.5)% in the common carp, *Cyprinus carpio* (1 fish was exposed to 100 cercariae). Recovery peak was found at week 16 post infection. After week 20 recovery rates were very low: around 0.1–0.2%.[9]

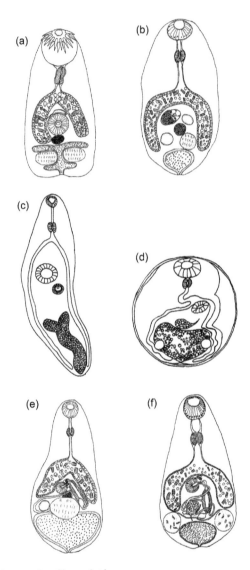

Figure 4.7 Excysted Metacercariae of heterophyids
a. *Centrocestus formosanus (original)*
b. *Haplorchis pumilio (original)*
c. *Heterophyopsis continua (after Komiya, 1965)[1]*
d. *Encysted metacercariae of Metagonimus yokogawai (after Komiya, 1965)[1]*
e. *Procerovum varium (original)*
f. *Stellantchasmus aspinosus (original)*

4.5.4.2 Infectivity of Metacercariae

In experiments, mature metacercariae of *O. viverrini* aged 5 to 20 weeks old were fed to golden hamsters, *Cricetus auratus* (30 metacercariae/hamster). Adult worm recovery rates were 21%

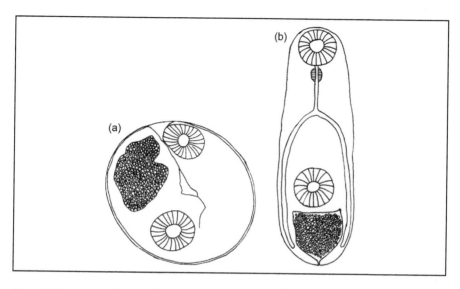

Figure 4.8 Metacercariae of opisthorchids (after Vajrastira et al., 1961)[8]
a. *Encysted metacercariae of Opisthorchis viverrini*
b. *Excysted metacercariae of Opisthorchis viverrini*

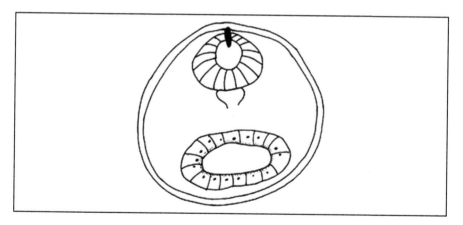

Figure 4.9 Encysted metacercariae of Nanophyetus salmincola (after Olsen, 1974)[11]

(15–40%), with recovery peak occurring in 12 week old metacercariae.[9]

4.5.5 Troglotrematidae

The microcercous cercariae enter the fish's blood vessels and migrate in fish before they encyst in muscles, gills, kidneys, and subcutaneous tissues[10] (Fig. 4.9).[11]

REFERENCES

1. Komiya Y. Metacercariae in Japan and adjacent territories. In: Morishita K, Komiya Y, Matsubayashi H, editors. *Progress of medical parasitology in Japan. VII.* Tokyo: Maruzen Co., Ltd; 1965. p. 1–309.

2. Caffara M, Gustinelli A, et al.. Morphological and molecular differentiation of *Clinostomum complanatum* and Clinostomum marginatum (Digenea: Clinostomidae). *J Parasitol* 2011;**97**:884–91.

3. Chen HT. Some early larval stages of *Centrocestus formosanus* (Nishigori, 1924). *Lingnan Sci J* 1948;**22**:93–103.

4. Martin WE, Kuntz RE. Some Egyptian heterophyid trematodes. *J Parasitol* 1955;**41**:374–82.

5. Lyster LL. Apophallus imperator sp. nov., a heterophyid encysted in trout, with contribution to its life history. *Can J Res* 1940;**18**:106–21.

6. Stunkard HW, Uzmann JR. The killifish, *Fundulus heteroclitus*, second intermediate host of the trematode, Ascocotyle (Phagicola) diminuta. *Biol Bull* 1955;**109**:475–83.

7. Martin WE. The life histories of some Hawaiin heterophyid trematodes. *J Parasitol* 1958;**44**:305–23.

8. Vajrastira S, Harinasuta C, Komiya Y. The morphology of the metacercaria of *Opisthorchis viverrini*, with special refence to the excretory system. *Ann Trop Med Parasitol* 1961;**55**:413–8.

9. Waikagul J. *The study on the infectivity of Opisthorchis viverrini metacercaria [M.Sc. Thesis].* Bangkok: Mahidol University; 1974. p. 1–70.

10. Schell SC. *How to know the trematodes.* Iowa: WMC Brown; 1970.

11. Olsen OW. *Animal parasites: their lifecycles and ecology.* 3rd ed. Baltimore: University Park Press; 1974. p. 564

CHAPTER 5

Molecular Systematics of Fish-Borne Trematodes

The fish-borne trematode (FBT) is an important group of parasites that can cause diseases, mostly notably among those belonging to the Opisthorchiidae, Heterophyidae, Clinostomidae, and Echinostomatidae families.[1] They include parasites of major medical significance, such as: *Opisthorchis* spp., *Clonorchis sinensis*, *Amphimerus* sp., *Metorchis* spp., *Haplorchis* spp. and *Metagonimus* spp. By morphology alone are sometimes remain many controversies to systemize the parasites. The problem on morphological identification can be occurred because of correspondence between organs arising from convergent evolution (homoplasy). The amount of homoplasy in existing data has frequently revealed inaccuracies in phylogenetic categorization of trematodes.[2,3] Currently, molecular information is available to clarify and confirm phylogenetic relationships and species diversity among the FBT.[4–10]

5.1 WHAT IS MOLECULAR SYSTEMATICS?

Systematics is the study of the diversification of living organisms from the past to the present to understand how any organism relates to other living things through time. During the past decade, the use of molecular data to infer the phylogenetic relationship among taxa has rapidly increased. Molecular tools have been developed and applied, such as DNA hybridization, polymerase chain reaction–restriction fragment length polymorphisms (PCR-RFLP), random amplified polymorphic DNA (RAPD), allozyme data, microsatellite DNA, and DNA sequences.[2,11] Currently, large numbers of DNA sequences are used as genetic markers, such as nuclear ribosomal genes, internal transcribed spacers and mitochondrial genes.[3,12,13]

Over the past decade, many major taxa of fish-borne trematodes have been clarified using phylogenetic relationships.[12] Molecular systematics has concentrated on two aims: to delineate common lineage and infer interrelationships, and to circumscribe species and strains,

Approaches to Research on the Systematics of Fish-Borne Trematodes. DOI: http://dx.doi.org/10.1016/B978-0-12-407720-1.00005-4

particularly in medically or zoonotically important trematodes.[13] Phylogenetic analysis has grown rapidly to assess the accuracy of hypotheses about biogeography, ecology, behavior, physiology, epidemiology, and almost every other aspect of biology.[14–16]

5.2 SPECIES IDENTIFICATION: MORPHOLOGICAL AND MOLECULAR APPROACHES

The species identification of digenean has mainly emphasized adult morphology and its host and geographic distribution.[3] However, morphology alone might be insufficient to identify species because of the small size of the adult stage and taxonomic characteristics combined with the invalidity of those characteristics.[17,18] There are issues of high-morphological similarities between closely related species: homoplasy,[2,19] phenotypic plasticity,[20] a lack of conserved structures,[21] and a lack of distinctive morphological characteristics.[22] The overestimation or underestimation of parasite diversity could easily have occurred. An expert is sometimes needed in order to achieve accuracy in identificationand to judge the variation among specimens, which can be problematic because of fixation in the staining process.

Currently, the molecular techniques, particularly those based on the polymerase chain reaction (PCR), have been used to test the hypotheses generated from the traditional method of species distinction. Nucleotide variation has been suggested as the best way to investigate inheritable differences.[23] The analysis of primary sequences reveals the potential of cryptic species complexes and their phylogeography, population genetics, and phylogenetic studies.[3] (The genomic DNA preparation and PCR methods are fully described in Appendix B)

5.3 ADVANTAGES AND DISADVANTAGES OF USING DNA SEQUENCES FOR SYSTEMATICS

There are four advantages to using DNA sequences in the analyses in systematics.[24,25] First, the DNA sequences used are selected from different genes or regions within genes that evolve at disparate rates. The regions, therefore, are able to be applied as genetic markers or DNA markers for a wide range of phylogenetic levels.[2,11] Second, the DNA marker selected is limited to only the nonrepetitive nucleotide positions in the genome. Third, the DNA sequence characters can resolve the problem of inheritable variation that an extensive problem in morphological characters.[26] Last, the techniques used in molecular work are rapidly developing and have now become available, reliable, and relatively inexpensive.

Disadvantages or limitations of DNA sequencing with respect to work on trematodes are mainly related to the limitations of sample collection. The lack of primary material, especially in the case of an inability to sample a wide range of sites throughout the geographic distribution of parasites, leads to incorrect identifications especially in the case of cryptic species, which is a technical problem in the PCR process producing contamination in the sample reactions. Molecular approaches do not always provide as much information as that gained from traditional approaches.

5.4 MOLECULAR MARKERS

A molecular marker (genetic marker) is a DNA fragment that is associated with a certain location within the genome.[27] The fragment selected should accurately determine phylogenetic relationships among species of interest at a particular category level. The utility of molecular markers can be specified by research ranging from the extreme micro- to macro-evolutionary levels.[28]

5.5 THE REGIONS OF DNA USED FOR MOLECULAR SYSTEMATICS

There are three DNA regions normally used as genetic markers in the phylogenetic study of metazoans: the nuclear coding region, ribosomal DNA, and genes in mitochondrial genomes.[28] Regions of DNA used as molecular markers are described in Table 5.1.

Table 5.1 Regions of DNA Used as Molecular, their Function and Genetic Variability for Molecular Systematics of Metazoan

Region	Function	Genetic Variability
Nuclear coding	Gene translated into proteins	Low-medium
Ribosomal DNA		
Non-coding	Non-functional RNA	High
rDNA gene	Produce RNA for protein synthesis process	Low-medium
Mitochondrial DNA	Involve the cellular energetic production	High

5.5.1 Nuclear Coding Region

The nuclear coding region is the region of a gene that produces a fully functional protein for an organism. The type of gene depends on its function within the genome. There are three recognized types of genes, including: protein-coding genes, RNA-specific genes, and regulatory genes. However, the regulatory gene is composed of untranscribed sequences. Only protein-coding and RNA-specific genes are able to produce functional proteins. Protein-coding genes are expressed under the transcription and translation processes. Some types of genes are found in all cell types and every life stage—such as some housekeeping genes—other genes only occur in a particular cell or tissue type—such as tissue-specific genes.[29] When utilizing genes as molecular markers, the housekeeping genes—such as the genes encoding glutamate dehydrogenase (GDH), heat shock proteins (HSP), and triose phosphate isomerase (TPI)—can be used to discriminate between species and sometimes on an intraspecific level. They can be used as markers because they are moderately conserved regions.[30]

5.5.2 Ribosomal DNA

The RNA-specific genes are usually similar in prokaryotes and eukaryotes. These genes are generally conserved and contained in a few sites of intron that must be spliced out before becoming functional RNA molecules.[29] The utilitization of these genes as molecular markers relates to the components of the gene making up the regions of genes and spacers.[3] In eukaryotic cells, nuclear DNA is tandemly organized with high copy numbers up to 5000 copies. Each repeating unit consists of genes coding for a small subunit (SSU), a large subunit (LSU), and 5.8 S rDNAs. The coding regions are separated by an internal transcribed spacer (ITS). The SSU and LSU regions are also separated by an external transcribed spacer (EST) and a nontranscribed spacer (NTS). The ETS and NTS are together called "an intergenic spacer (IGS)." The 5.8 S rDNA is embedded in both of the two ITS's (ITS1 and ITS2) (Fig. 5.1).[3]

The ribosomal genes and ITS regions of rDNA have the potential to be molecular markers for molecular systematics in different prospective works. Generally, the 18 S rDNA gene is the slowest to evolve sequences used to infer deep phylogenetic relationships, such as family category level. The 28 S rDNA gene has a larger and faster rate of nucleotide evolution than the 18 S rDNA gene.[3] However, both

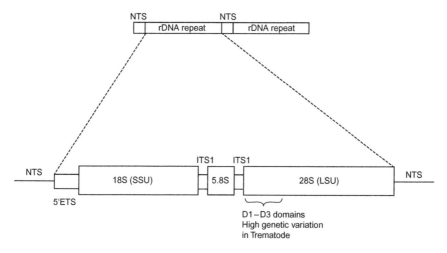

Figure 5.1 The ribosomal RNA gene of eukaryotes. (A) Tandems of rRNA gene clusters. There is a very large rDNA array and low polymorphism between rDNA repeat unit. (B) The gene contains 18 S or small subunit (SSU), 5.8 S, and 28 S or large subunit (LSU) tracts. D1 to D3 domain is expansion segment of 28 S rRNA gene, which has high genetic variation. NTS is nontranscribed spacer, ITS is internal transcribed spacer having 2 segments numbered from 5' end (so-called ITS1 and ITS2, respectively). ETS is external transcribed spacer.

regions have been used to study genetic differentiation within digenean families.[31] The 5.8 S rDNA gene has a similar level of gene conservation to the 18 S rDNA gene, although it is between the ITS1 and ITS2 regions. However, it is too short to use alone in molecular phylogenetic analysis.[32]

In the digenean subclass, the ITS1 and ITS2 regions are not similar in their rates of evolution. The ITS1 region seems more conserved than the ITS2 region. The nature of the ITS1 rDNA spacer allows for the characterization of the digenean at different category levels. The 5' end of ITS1 has a higher variability than the 3' end, which allows for discriminating among species, while the 3' end has a lower variability than the 5' end, so it can provide information on the relative systematics position of the 3' end.[3]

A limitation in the use of ITS1 in molecular systematics studies of digeneans is the presence of tandem repeats. Most ITS1 regions are composed of three repeating elements: first, at the short 5' end region; second, within a tract of ITS1 containing more than 2 repeating parts; third, at the 3' region, which usually does not repeat, but presents within ITS1 of *Schistosoma* spp.[33] Therefore, the number of nucleotides constituting a repeat and the number of copies of repeats present vary in the species and family.[34,35]

The ITS2 region of rDNA is generally used in molecular systematics. It contains variable sites more than the ITS1 region and does not contain a repeat element.[33] It is highly variable in length within and between families, however, it is relatively high conserved at the species level. Thus, it might not be suitable for studying the molecular systematics above the genus level.[3]

There are limitations of the use of ITS in the molecular systematics of digenean species in the determination of species. Genetic differentiation at the species level is sometimes not clear when morphology is compared. Problematic issues for using ITS regions as molecular markers are: 1) some species have identical sequences to those of other closely related species, 2) some species have intraspecific variation.[3]

5.5.3 Mitochondrial DNA

Because of its high evolutionary rate, animal mitochondrial DNA has become a popular molecular marker for molecular systematics, not only at the species level, but also at the intraspecific level.[36] Many mtDNA sequences are available in the GenBank database, including those of helminths. Many previous works reported genetic variability in many helminthic genera, such as: *Ascaris*, *Onchocerca*, *Schistosoma*, *Fasciola*, *Paragonimus*, *Echinostoma*, *Echinococcus*, *Taenia*, *Opisthorchis*, and *Haplorchis*, among others.[37−43]

In vertebrates, mtDNA is presented in multiple copies, usually 1,000−10,000 copies per cell, and is expressed in about 13 different proteins of enzyme complexes in the respiratory chain supporting ATP-synthesis.[36] It is hypothesized that the mitochondrial genome is a symbiont of eukaryotic cells. This genome is clonal and never undergoes recombination. The mtDNA is generally inherited maternally, which is a limitation for using mtDNA in molecular systematics.[36] The genetic evolution record never comes from the male. The backcrossing of female interspecific hybrids to males can not occur.[44] The lack of male lineage leads to the loss of genetic data for actual discussion in very closely related species and species complexes.

All molecular markers reviewed have certain limitation(s). Therefore, using only one kind of marker leads to misunderstanding of and/or loss of the entire genetic relationships of the group of species of interest.

5.6 MOLECULAR SYSTEMATICS OF FISH-BORNE TREMATODES

In the group of fish-borne trematodes, there are four major families containing species recognized to be zoonotic parasites, such as: Opisthorchiidae, Heterophyidae, Echinostomatidae, and Clinostomidae.[1] The main aim in molecular technique is confirming morphological identification, especially through DNA sequence analysis and inferring the phylogenetic relationships to clarify the generic affiliation of all species/ specimens identified.

Previous studies have revealed the phylogenetic positions of those four families. The Opisthorchiidae has a close relationship with Heterophyidae and they belong to the superfamily Opisthorchioidea. Echinostomatidae is a family of parasite that is associated with a broad range of definitive hosts and has a wide geographical distribution. Among the genera in Echinostomatidae, the DNA sequence data of *Echinochasmus*[12] in fish is very limited in DataBank; this family is not mentioned with respect to molecular systematics in this chapter. Molecular markers for studying the molecular systematics and molecular identification of species in families Opisthorchiidae, Heterophyidae, and Clinostomidae are provided in Appendix B, Table 1.

5.6.1 Opisthorchiidae Looss 1899 and Heterophyidae Leiper 1909

Opisthorchiidae and Heterophyidae belong to the superfamily Opisthorchioidea Looss, 1899, which also has Cryptogonimidae as a member. These three families are similar in their morphology and life-cycle.[45] Members of the Opisthorchiidae family include a liver fluke—consisting of over 30 genera—and several important human pathogens—such as *Opisthorchis viverrini*, *O. felineus*, and *Clonorchis sinensis*. The Heterophyidae family is a family of intestinal flukes that consists of over 50 genera, including some important human pathogens, such as *Haplorchis* spp. and *Metagonimus* spp.[46,47] Among the species in superfamily Opisthorchioidea, there are similar fundamental morphologies and only subtle differences in body structures. Taxonomic classification of this superfamily has been reassessed according to those characteristics.[45,48] The uncertain relationships among genera and species within Opisthorchioidea were indicated based on morphological characteristics.[49] Molecular characteristics have been used to indicate the classification of the digenean based on small and large nuclear ribosomal subunits. The Phylogenetic tree infers Opisthorchiidae and Heterophyidae have a close relationship.[12]

The relationship between these two families was confirmed by Thaenkham et al. (2012) using all available data in GenBank, such as 18 ribosomal DNA and internal transcribed spacer (ITS2) sequences. The results suggest that the Opisthorchiidae and Heterophyidae families are inseparable. However, this study suggested that ITS2 sequences could not provide high confident phylogenetic relationships[8]. To confirm paraphyletic relationship between those two families was performed successfully by the combined sequences of 18 S and 28 S ribosomal DNA.[6] Fig. 5.2 shows heterophyid intestinal flukes *Euryhelmis costaricensis* and *Cryptocotyle lingue* are grouped together in the same clade with Opisthorchiid liver flukes *Amphimerus ovalis*, *Opisthorchis viverrini, and Clonorchis sinensis*, with high bootstrap support. By morphology, the five species just described share the characteristic of the presence of lateral vitelline follicles. From traditional taxonomy, the worms of Opisthorchiidae and Heterophyidae are classified by the presence or absence of a genital sac/ventrogenital sac.

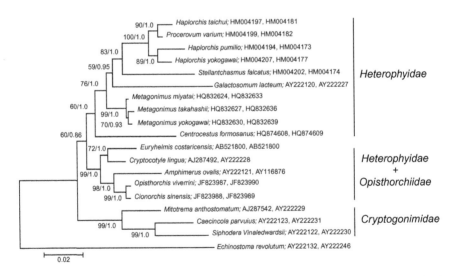

Figure 5.2 Phylogenetic tree inferred from concatenated SSU and LSU sequences of superfamily Opisthorchioidea. Each node is supported by bootstrap values from maximum likelihood analysis (using MEGA 5.0 program) and posterior probabilities from Bayesian inference (using MrBayes 3.1 program). Echinostoma revolutum is an outgroup. DNA sequences were analyzed and the tree was constructed under the GTR + G + I model (The best-fit model of nucleotide substitution; GTR: General Time Reversible, G: gamma distribution, I: invariable site).

The commonalities between these two families discovered through molecular systematics suggests that the key morphological characteristics traditionally used may not be appropriate. Opisthorchiid liver flukes and Heterophyid intestinal flukes share some second intermediate hosts (freshwater fish) and definitive hosts (cat, dog, fish-eating bird).[50,51] This evidence suggests that these two families may have a closer common evolution than the Cryptogonimidae family, which is generally found in marine fish hosts.[45] Based on morphology, the adult Opisthorchiid liver fluke and adult Heterophyid intestinal fluke are very fdifferent, however egg and larval stages (cercariae and metacercariae) are similar. Egg and larvae stages are an important key for the taxonomy of digenean parasites.[6]

5.6.2 Clinostomidae Lühe, 1901

Clinostomidae is a family of fish-borne trematodes belonging in the subfamily Clinostomatinae Luhe, 1901. *Clinostomum* Leidy, 1856 have been unstable in taxonomic history and species composition because there is a high degree of morphological variability within the species.[52,53] This genus required taxonomic revision. There are 13 valid species that were recognized.[54] *Clinostomum complanatum* Rudolphi, 1814 has been reported to be a human pathogen.[55] This species has been confused in taxonomic identification because of a lack of information of its taxonomic status by linking the larval and adult stages and also confusing its geographical distribution. *C. complanatum* and *C. marginatum* are the species that have been most problematic in their morphology. The adult worms of these two species show high morphological similarity. *C. complanatum* is located throughout Asia, Europe, and Africa, in the so-called old world. and is thus called the"European type." *C. marginatum* is found in North America in the so-called new world, and is called the"American type."[56] Several authors considered these two species to be the same species based on the morphological character of the adult worm and classified the *C. complanatum* as a synonym of *C. marginatum*.

Therefore, to strongly confirm taxonomic level, research would require sample collection from a wide geographic and host range and would also require DNA sequence data to confirm the validity of taxonomic status.

Dzikowski et al. (2004) reported the potential of a small subunit ribosomal DNA (18 S rDNA). *C. complanatum* from Israel and *C. marginatum*

from the United States were analyzed using the size of their PCR ampli-
cons (mainly between 670 and 740 bp). The 18 rDNA sequences quickly
revealed that these two species are genetically different.[56] Based on
molecular results, the morphological differences between *C. complanatum* and *C. marginatum* suggest that they should be classified by the
position of gonotyle and the shape of the vitelline glands.[56–58]
However, the utility of 18SrDNA was reduced by the evolutionary conservation. This marker could not be used to discriminate between
C. cutaneum Paperna, 1964 and *C. phalacrocoracis* Dubois, 1930, which
were obviously separated by morphology.[54] Therefore, to resolve the
uncertain genetic relationship between *C. complanatum* and *C. marginatum*, additional molecular markers were required and then the result
obtained was linked to morphological information.[59]

Caffara et al. (2011) used sequences of ITS regions (ITS1 and
ITS2), which are recommended as potential markers for taxonomic
studies at the species level[3] and also used the barcoding region of cytochrome c oxidase subunit 1 (COI)—developed by Bowles et al.
(1995)—as molecular markers.[60] The DNA sequence data analyzed
then clarified the morphological characteristics that could be the
potential characteristics for discriminating between *C. complanatum*
and *C. marginatum*.[9] The study indicated reliable morphological differences in the genital complex at the metacercarial and adult stages of
both species. The morphometric study suggested that the distance
between the suckers and the body width of the metacercariae may be
useful for species discrimination. The morphological differences were
supported by the result of phylogenetic analyses from ITS sequences
(Fig. 5.3). This study confirmed that *C. complanatum* and *C. marginatum* have a distinct genetic separation.[9]

Sereno-Uribe et al. (2013) revealed the phylogenetic relationship
among *Clinostomum* spp. by collecting definitive and intermediate
hosts from 18 localities across Mexico (across the Nearctic and across
the Neotropical biogeographical regions) and by then using mitochondrial and ribosomal DNA sequences. The combination of molecular
data with morphology, host association, and geographical distribution
indicated the potential recognition of the *Clinostomum* cryptic species.[10] This study reported a new species named *C. tataxumui* n. sp.
This species was described based on adult worms found in the mouth
cavity of fish-eating birds and also metacercariae found in freshwater

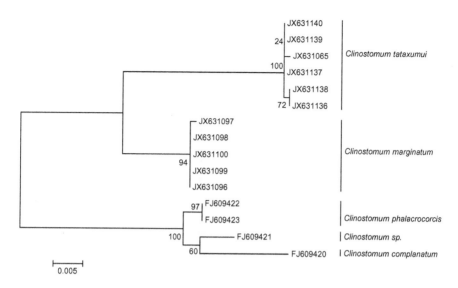

Figure 5.3 Midpoint rooting of the phylogenetic tree inferred from ITS sequences of Clinostomum spp. Each node supported by bootstrap value from maximum likelihood analysis using MEGA 5.0 program. DNA sequences were analyzed and the tree was constructed under the K2 + G model.

and estuarine fish. This species has a few morphological characteristics similar to two other congeners found in Mexico, namely *C. intermedialis* and *C. complanatum*.[61] Based on samples collected in the study, there was no specimen of *C. complanatum*, but there was *C. margunatum*. The phylogenetic analysis (Fig. 5.4), genetic divergence, and morphometric results confirmed *C. tataxumui* is the new valid species and *C. complanatum* is not currently found in Mexico.[10] This study requires more thoughtful sampling in the areas—at least in Mexico—however, the information obtained supports the hypothesis that *C. marginatum* is the American form.

5.7 SUMMARY AND FUTURE DIRECTION

Many decades ago, systematics of fish-borne trematodes was based only on morphological characteristics that sometimes made for confusion in the taxonomic status because of unclear or unstable morphological diagnostic characteristics. Currently, genetic markers have been selected from DNA regions that can be used to identify individuals or species according to the DNA sequence variation observed. The phylogenetic relationships of fish-borne trematodes in the Opisthorchiidae, Heterophyidae, and Clinostomidae families are

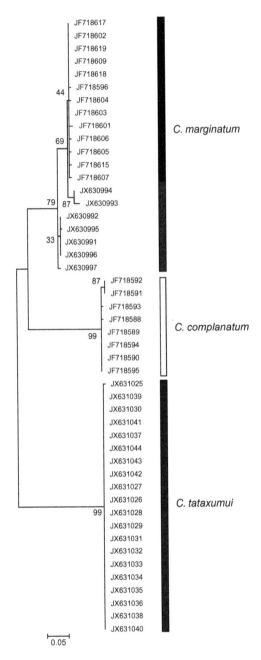

Figure 5.4 Midpoint rooting phylogenetic tree inferred from COI sequences of Clinostomum spp. Each node supported by bootstrap value from maximum likelihood analysis using MEGA 5.0 program. DNA sequences were analyzed and the tree was constructed under the HKY + I model. Black bar stands for samples collected from Canada, gray bar stands for samples collected from Mexico, and white bar stands for samples collected from Italy.

consistent with some of the criterial diagnostic characteristics used to classify at the family and congeneric levels, whereas some criterial morphological characteristics used did not link with their phylogenetic positions. Therefore, the molecular data revealed that morphological characteristics may not be enough for accuracy in classification. The data indicated reliable morphological differences and consistent characteristics through evolutionary time. Morphological characteristics for discriminating species could require further taxonomic revision. However, the FBT group is the group of parasites that have a high degree of diversity because they are distributed widely across various geographical areas and have a wide range of hosts throughout complex life cycles. Practical sample designs to collect the parasites has been required to gain more information about FBTs, particularly in fish intermediate hosts and definitive hosts for confirming taxonomic status by linking the larvae with adults found in the same locality.

REFERENCES

1. Sohn W-M. Fish-borne zoonotic trematode metacercariae in the republic of Korea. *Korean J Parasitol* 2009;**47**:S103—13.

2. Blair D. Use of molecular data in trematode systematics: strengths and weakness. *Acta Parasitol Turcica* 1996;**20**:375—86.

3. Nolan J, Cribb TH. The use and implications of ribosomal DNA sequencing for the discrimination of Digenean species. *Adv Parasitol* 2005;**60**:101—63.

4. Dzikowski R, Levy MG, Poore MF, et al. Use of rDNA polymorphism for identification of Heterophyidae infecting freshwater fishes. *Dis Aquat Org* 2004;**59**:35—41.

5. Thaenkham U, Dekumyoy P, Komalamisra, et al. Systematics of the subfamily Haplorchiinae (Trematoda: Heterophyidae), based on nuclear ribosomal DNA genes and ITS region. *Parasitol Int* 2010;**59**:460—5.

6. Thaenkham U, Nawa Y, Blair D, et al. Confirmation of the paraphyletic relationship between Opisthorchiidae and Heterphyidae using small and large subunit ribosomal DNA sequences. *Parasitol Int* 2011;**60**:521—3.

7. Thaenkham U, Nuantanong S, Vonghachack Y, et al. Discovery of Opisthorchis lobatus (Trematoda: Opisthorchiidae): a new record of small liver flukes in the Greater Mekong Subregion. *J Parasitol* 2011;**97**:1152 -8.

8. Thaenkham U, Blair D, Nawa Y, et al. Families Opisthorchiidae and Heterophyidae: are they distinct? *Parasitol Int* 2012;**61**:90—3.

9. Caffara M, Locke SA, Gustinelli A, et al. Morphological and molecular differentiation of Clinostomum complanatum and Clinostomum marginatum (Digenea: Clinostomidae) metacercariae and adults. *J Parasitol* 2011;**97**:884—91.

10. Serono-Uribe A, Pinacho-Pinacho CD, García-Varela M, et al. Using mitochondrial and ribosomal DNA sequences to test the taxonomic validity of Clinostomum complanatum Rudolphi, 1814 in fish-eating birds and freshwater fishes in Mexico, with the description of the new species. *Parasiol Res* 2013;**112**:2855—70.

11. Hwang U-W, Kim W. General properties and phylogenetic utilities of nuclear ribosomal DNA and mitochondrial DNA commonly used in molecular systematics. *Korean J Parasitol* 1999;**37**:215–28.

12. Olsen PD, Cribb TH, Tkach VV, et al. Phylogeny and classification of digenean (Plathyhelminthes: Trematoda). *Int J Parasitol* 2003;**33**:733–55.

13. Olsen PD, Tkack VV. Advance and trends in the molecular systematics of the parasitic platyhelminthes. *Adv Parasitol* 2005;**60**:165–243.

14. Mantooth SJ, Riddle BR. Molecular biogeography: the intersection between geographic and molecular variation. *Geogr Comp* 2011;**5**:1–20.

15. Garland T, Bennett AF, Rezende EL. Phylogenetic approaches in comparative physiology. *J Exp Biol* 2005;**208**:3015–35.

16. Brooks DR, McLennan DA, Carpenter JM, et al. Systematics, ecology, and behavior: Integrating phylogenetic patterns and evolutionary mechanism. *BioScience* 1995;**45**:687–95.

17. Schulenburg JHG, Englisch U, Waegele JW. Evolution of ITS1 rDSNA in the Digenea (Platyhelminthes: Trematoda): 3' end sequence conservation and its phylogenetic utility. *J Mol Evol* 1999;**48**:2–12.

18. Maldonado JA, Loker ES, Morgan JAT, et al. Description of the adult works of a new Brazilian isolation of Echinostoma paraensei (Platyhelminthes: Digenea) from its natural vertebrate host Nectomys squamipes by light and scanning electron microscopy and molecular analysis. *Parasitol Res* 2001;**87**:840–8.

19. Monis PT. The importance of systematics in parasitological research. *Int J Parasitol* 1999;**29**:381–8.

20. Galazzo DE, Dayanandan S, Marcogliese DJ, et al. Molecular systematics of some North American species of Diplostomum (Digenea) based on rDNA-sequence data and comparisons with European congeners. *Can J Zool* 2002;**80**:2207–17.

21. Jousson O, Bartoli P. Molecules, morphology and morphometrics of Cainocreadium labracis and Cainocreadium dentecis n. sp. (Digenea: Opecoelidae) parasitic in marine fishes. *Int J Parasitol* 2001;**31**:706–14.

22. Bartoli P, Jousson O. The life-cycle of Podocotyle scorpaenae (Opecoelidae: Digenea) demonstrated by comparative analysis of ribosomal DNA sequences. In: Combes C, Jourdance J, editors. *Taxonomy, ecology and evolution pf metazoan parasites*. Perpignan: Presses Universitaires de Perpignan; 2003. p. 17–28.

23. Morgan JAT, Blair D. Mitochondrial ND1 gene sequences used to identify echinostome isolate from Australia and New Zealand. *Int J Parasitol* 1999;**28**:498–502.

24. Nadler SA. Molecular approaches to studying helminth population genetics and phylogeny. *Int J Parasitol* 1990;**20**:11–29.

25. McManus DP, Bowel J. Molecular genetic approaches to parasite identification: their value in diagnostic Parasitology and systematics. *Int J Parasitol* 1996;**26**:687–704.

26. Hillis DM. Molecular versus morphological approaches to systematics. *Ann Rev Ecol Syst* 1987;**18**:23–42.

27. http://en.wikipedia.org/wiki/Molecular_marker

28. Avise JC. *Introduction. Molecular markers, natural history and evolution*. Chapman and Hall; 1994. pp. 3–15

29. Li WH, Graur D. *Gene structure and mutation. Fundamentals of molecular evolution*. 2nd ed. Sunderland, Massachusetts, USA: Sinauer Associates Inc.; 1991. pp. 5–38

30. Monis PT, Giglio S, Keegan AR, et al. Emerging technologies for the detection and genetic characterization of parasites. *Trends Parasitol* 2005;**21**:340–6.

31. Kaukas A, Rollinson D. Interspecific variation within the 'hypervariable' region of 18S ribosomal RNA gene among species of Schistosoma Weinland, 1858 (Digenea). *Syst Parasitol* 1997;**36**:157–60.

32. Hillis DM, Dixon MT. Ribosomal DNA: molecular evolution and phylogenetic inference. *Q Rev Biol* 1991;**66**:411–53.

33. Kane RA, Rollinson D. Repetitive sequences in the ribosomal DNA internal transcribed spacer of Schistosoma haematobium, Schistosoma intercalatum, and Schistosoma mattheei. *Mol Biochem Parasitol* 1994;**63**:153–6.

34. Van Herwerden, Blair D, Agatsuma T. Intra- and Inter-specific variation in nuclear ribosomal internal transcribed spacer 1 of the Schistosoma japonicum species complex. *Parasitol* 1998;**116**:311–7.

35. Van Herwerden, Blair D, Agatsuma T. Intra- and inter-individual variation in ITS1 of Paragonimus westermani (Trematoda: Diagenea) and related species: implications for phylogenetic studies. *Mol Phylo Evol* 1999;**12**:67–73.

36. Le TH, Blair D, Donald P, et al. Mitochondrial genomes of human helminths and their use as markers in population genetics and phylogeny. *Acta Trop* 2000;**77**:243–56.

37. Unnasch TR, Wiliams SA. The genomes of Onchocerca volvulus. *Int J Parasitol* 2000;**30**:543–52.

38. McMamus DP, Bowles J. Molecular genetic approaches to parasite identification, their value in applied parasitology and systematics. *Int J Parasitol* 1996;**26**:687–704.

39. Morgan JAT, Blair D. Relative merits of nuclear ribosomal internal transcribed spacers and mitochondrial CO1 and ND1 genes for distinguishing among Echinostoma species. *Parasitol* 1998;**116**:289–97.

40. Morgan JAT, Blair D. Mitochondrial ND1 sequences used to identify echinostome isolates from Australia and New Zealand. *Int J Parasitol* 1998;**28**:493–502.

41. Bowles J, McManus DP. NADH dehydrogenase 1 gene sequences compared for species and strains of the genus Echinococcus. *Int J Parasitol* 1993;**23**:969–72.

42. Bowles J, McManus DP. Genetic characterisation of the Asian Taenia, a newly describes taeniid cestode of humans. *Am J Trop Med Hyg* 1994;**50**:33–44.

43. Gasser RB, Zhu Z, McManus DP. NADH dehydrogenase subunit 1 sequence compared for nine members of the genus Taenia (Cestoda). *Int J Parasitol* 1999;**29**:1965–70.

44. Willson AC, Cann RL, Carr SM, et al. Mitochondrial DNA and two perspectives on evolutionary genetics. *Biol J Linn Soc* 1985;**26**:375–400.

45. Bray RA. Superfamily Opisthorchioidea looss, 1899. In: Bray RA, Gibson DI, Jones A, editors. *Keys to the trematoda*, vol. 3. UK: CAB International and Natural History Museum; 2008.

46. Chai JY, Murrel D, Lymbery AJ. Fish-borne zoonoses:status and issues. *Int J Parasitol* 2005;**35**:1233–54.

47. Keiser J, Utzinger J. Food-borne trematodiases. *Clin Microbiol Rev* 2009;**22**:466–83.

48. Price EW. A review of the trematode superfamily Opisthorchioidea. *Proc Helminthol Soc Washington* 1940;**7**:1–13.

49. Scholz T. Family Opisthorchiidae Looss, 1899. In: Bray RA, Gibson DI, Jones A, editors. *Keys to the trematoda*, vol. 3. UK: CAB International and Natural History Museum; 2008.

50. Kaewkes S. Taxonomy and biology of liver flukes. *Acta Trop* 2003;**88**:177–86.

51. Namue C, Rojanapaibul A, Wongsawad C. Occurrence of two heterophyid metacercariae Haplorchis and Haplorchoides in cyprinoid fish of some districts in Chiang Mai and Lumphun Province. *Southeast Asian J Trop Med Public Health* 1998;**29**:401–5.

52. Yamaguti S. *Synopsis of digenetic trematodes of vertebrates*, vol. 1. Tokyo: Keigaku Publishing Co.; 1971. pp. 1074

53. Matthews D, Cribb TH. Digenetic trematodes of the genus Climostomum Leidy, 1856 (Digenea: Clinostomidae) from birds of Queenland, Australia, including C. wilsoni n. sp. from Egretta intermedia. *Syst Parasitol* 1998;**39**:199–208.

54. Gustinelli A, Caffara M, Florio D, Otachi EO, et al. First description of the adult stage of Clinostomum cutaneum Paperna, 1964 (Digenea: Clinostomidae) from grey herons Ardea cinerea L. and a redescription of the metacercaria from the Nile tilapia Oreochromis niloticus (L.) in Kenya. *Syst Parasitol* 2010;**76**:39–51.

55. Tiewchaloern S, Udomkijdecha S, Suvouttho S, et al. Clinostomum trematode from human eye. *Southeast Asian J Trop Med Public Health* 1999;**30**:382–4.

56. Dzikowski RMG, Levy MF, Poore JR, et al. Clinostomum complanatum and Clinostomum marginatum (Rudolphi, 1819) (Digenea: Clinostomidae) are separate species based on differences in ribosomal DNA. *J Parasitol* 2004;**90**:413–4.

57. Lo CH, Haber WF, Kou GH. The study of Clinostomum complanatum (Rudolphi, 1814) II. The life cycle of Clinostomum complanatum. CAPD Fisheries Series No. 8. *Fish Dis Res* 1981;**4**:26–56.

58. Finkelman S. *Infections of Clinostomoidea in the sea of Galilee fish. M.Sc. Thesis.* Jerusalem. Israel: Hebrew University of Jerusalem; 1988.

59. Perkins SL, Martinsen ES, Falk BG. Do molecules matter more than morphology? Promises and pitfalls in parasites. *Parasitol* 2011;**138**:1664–74.

60. Bowles JD, Blair D, McManus DP. A molecular phylogeny of the human schistosomes. *Mol Phylo Evol* 1995;**4**:103–9.

61. Pérez-Ponce de León G, García Prieto L, Mendoza Garfias B. *Trematode parasites (Platyhelminthes) of wildlife vertebrates in Mexico. Zootaxa 1534.* Auckland. New Zealand: Magnolia Press; 2007.

62. Tkach VV, Pawlowski J, Sharpilo VP. Molecular and morphological differentiation between species of Plagiorchis vespertilionis group (Digenea, Plagiorchiidae) occurring in European bats, with a re-description of P. vespertilionis (Muller, 1780). *Syst Parasitol* 2000;**47**:9–22.

CHAPTER 6

Methods of Molecular Study: DNA Sequence and Phylogenetic Analyses

The goal of molecular systematics is a general study of the phylogenetic relationships of living organisms. The DNA regions called "genetic markers" have been developed to reveal evolutionary relationships and also to identify their phylogenetic positions. Although any part of the genome of an organism is able to provide some information about taxonomic affiliation, some regions are more useful than others.[1] The regions that are currently available for molecular systematics are those encoding the ribosomal small and large subunit sequences, the internal transcribed spacers of rRNA, and the mitochondrial genes.[2-5] The characteristics of regions used for molecular systematics were described in Chapter 5. The genetic markers should be chosen based on the research purpose because each marker has limitations and is powerful for a specific hierarchy in the evolutionary range.[6,7] The pitfall of wrong marker selection is the wrong interpretation of phylogenetic relationships. If the marker used detects too much variation, it is possible that the samples are too different (homoplasy). If the marker detects too little variation, it means that the marker used cannot detect genetic differences between samples.[8] Moreover, using only a single genetic marker is not enough to indicate the most detail possible about phylogenetic relationships because of the limitations of each genetic marker. To reveal a reliable result, both nuclear DNA and mitochondrial genes are always used to support each other.

6.1 CRITERION FOR SELECTING APPROPRIATE GENETIC MARKERS

Based on general properties and phylogenetic utilities, the nuclear ribosomal DNA and mitochondrial DNA are powerful markers for molecular systematics.[2] In the past decade, indirect methods of composing DNA hybridization, polymerase chain reaction—restriction fragment length polymorphisms (PCR-RFLP), random amplified polymorphic DNA (RAPD), and allozyme data were the preferred genetic markers

Approaches to Research on the Systematics of Fish-Borne Trematodes. DOI: http://dx.doi.org/10.1016/B978-0-12-407720-1.00006-6

for resolving phylogenetic questions because the materials used were inexpensive. However, those methods are not able to estimate genetic variations of the specific DNA region among taxa examined.[9] In years of the post-genomic era, the costs of PCR amplification and DNA sequencing have decreased. Researchers can obtain DNA sequences from organisms of interest within a few days. Moreover, there has been a rapid increase in DNA sequence information in DNA data banks such as DDBJ (DNA Data Bank of Japan), ENA (European Nucleotide Archive), and GenBank.

Criteria for selecting the appropriate genetic marker have to be concerned with the systematic questions and hierarchical taxonomic levels. This is the most critical step for phylogenetic analysis because the application of an inappropriate molecular marker leads to the misinterpretation of phylogenetic relationships. A lack of understanding of the properties of molecular markers suitable for categorical levels examined is a major problem.

The genetic markers for molecular systematics are generally the gene or DNA regions of nuclear ribosomal rRNA and the genes of the mitochondrial genome. For studies of phylogenetic relationships among species from different genera to families, the highly conserved regions of rDNA are useful, such as the entire small subunit ribosomal RNA gene (ssrDNA; 18 S rDNA) and the partial (D1-D3) large subunit ribosomal RNA gene (lsrDNA; 28 S rDNA) sequences. These regions were used to estimate the phylogeny of 163 digenean taxa from 77 nominal families.[4] Using only part of the 18 S rDNA gene can cause a gap in phylogenetic information because over half of its total length is very conserved regions.[9]

For the 28 S rDNA gene, the 5′ end of this gene is known as the variable domain and has been used for studying the deep branches of the metazoans and some species of digeneans.[10,11] However, Shylla et al. (2013) studied the species discrimination of the paramphistomes (Trematoda: Digenea) and recommended the use of the D1-D3 28 S rDNA for a significant resolution of the taxa corroborating with the taxonomy of the flukes. Although the D1 region contains the most variable site, the D2 region is the most robust domain, comprising compensatory mutation in the helices of its structural constraints.[12] The data of these regions is suitable for inference of phylogenetic relationships among species, genera, and closely related families.[9,11,12]

The internal transcribed spacer region of ribosomal DNA (ITS rDNA) has been used as the default marker for phylogenetic analysis. The ITS rDNA is the potential DNA marker to confirm the distinctness of valid trematode species and also to reveal the existence of cryptic species.[5] Nolan and Cribb (2005) reviewed 63 studies that used the entire ITS region. The ITS1 region is suggested as providing greater resolution of species differentiation than the ITS2 because of the presence of variable repeat units. They are low or apparently absent of intraspecific variation. Therefore, this region might not be a suitable marker for inter-population study. Morgan and Blair (1995) suggested that the ITS2 region might be too conserved among closely related species. The ITS1 region is the greatest variable site of the ITS and has the potential to identify variation that is not detected by ITS2. Using the entire ITS region has been recommended to resolve the problem of distinguishing among the digenean species, although ITS2 has successfully been applied for discriminating species from many digenean families.[13] A very low variation of ITS2 is required to prove that it would be the good marker for molecular identification.[5]

The mitochondrial genome is a powerful source of genetic markers for the phylogenetic study of closely related species/ complex species and inter-populations because inter- and intra-specific variations are to be expected.[9,14–16] The recommended mitochondrial genes are listed in Table 6.1. However, the most popular marker used for phylogenetic studies of fish-borne trematodes are the *cox1* and *nad1* genes.[17–19] Recently, there are many mitochondrial genomes of human helminths

Table 6.1 Recommended Mitochondrial Genes and their Products Utilized as Potential Molecular Markers

Gene	Gene Product	Gene used for fish-borne Trematode Research
*cox*1,2,3	COX1,2,3	*cox*1[17]
*nad*1,2,3,4 L,4,5,6	NAD1,2,3,4 L,4,5,6	*nad*1[18]
cob	COB	
*atp*6,8	ATP6,8	
*rrn*L	LrRNA	
*rrn*S	SrRNA	
*trn*M, *trn*W, etc.	tRNA (M), tRNA (W), etc.	

Note: cox: *Cytochrome c oxidase genes;* nad: *mitochondrially encoded NADH dehydrogenase;* cob: *mitochondrially encoded cytochrome b;* atp: *mitochondrially encoded ATP synthase;* rrnL: *large ribosomal RNA gene;* rrnS: *small ribosomal RNA gene;* trn: *transfer ribosomal RNA gene*

that have been completed; this will facilitate the design of universal primers for applying with the various mitochondrial genes in many taxa. This is because unequal DNA variation among mitochondrial genes must be addressed.[3]

Different selective forces lead to the evolution of various DNA/gene regions in genomes with varying degrees of sequence conservation. Therefore, the appropriate DNA regions should be considered carefully before selection as the molecular markers for resolving the addressed question on phylogenetic relationships with respect to various taxonomic hierarchies. The utilities of each molecular marker for the appropriate taxonomic hierarchy are summarized in Table 6.2.

6.2 PHYLOGENETIC ANALYSIS

Phylogenetics is the study of evolutionary relationships by inferring or estimating the evolutionary past. Based on DNA or protein sequences, the evolutionary relationship can be described through molecular phylogeny, which became an achievable method for researchers in the genomic era.[20]

The evolutionary history is inferred from phylogenetic analysis and is depicted as a treelike diagram, which represents an estimated pedigree of the inherited relationships among multigene families (gene tree) or among a single gene from many taxa (species tree).[20,21] The internal nodes of the gene tree correspond to gene duplication, while the nodes of a species tree harmonize with speciation events. All parts of the tree illustrated represent evolutionary relationships and also provide a historical pattern of ancestry, divergence, and descent. The trees are

Table 6.2 Utility of each Molecular Marker for Applying to the Appropriate Taxonomic Hierarchy

Categorical Level	Level of DNA Variability	Example of DNA Marker Used
Interspecific		
Close species/complex	ca. 1%/my	mtDNA, ITS rDNA
Different genera to families	ca. 0.1%/my	Some LSU rDNA/ SSU rDNA
Different classes to phyla	ca. 0.1%/my	D1 of LSU rDNA, SSU rDNA
Inter-population		
Phylogeography (population structure)	Medium to high	mtDNA

depicted as branch tips (terminal nodes) that join at internal nodes that represent inferred speciation events. The lineage splitting represents descendant sister groups and common ancestors from two or more related lineages.[22] The pattern of tree branching is called "topology" (Fig. 6.1). From the internal node, the grouping of operational taxonomic units or OTU's is a clade or a "monophyletic group." The monophyletic group represents all members that share or are derived from a unique common ancestor. A group that is excluded from its descendantsis called aparaphyletic group. A polyphyletic group is a taxonomic group that includes members (as genera or species) from different ancestral lineage.

Molecular phylogenetic trees are drawn with proportional branch lengths corresponding to the amount of evolution between connecting nodes. The widths of the nodes have no meaning in terms of evolution, but adjust to the space between branches. All of the branches can rotate freely around their nodes.[21]

The most general types of phylogenetic trees are rooted and unrooted trees. A rooted tree is a directed tree with a unique node

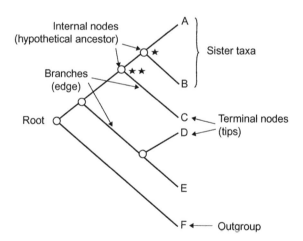

Figure 6.1 The tree terminology. From the right, A to F are the terminal nodes or "tips" of the tree representing individual species which are modern species. The terminal nodes are connected to one another through branches that join at internal nodes. Internal node represents hypothetical ancestor and infers lineage splitting or speciation events. Branches defines the relationship between the taxa in term of descend and ancestor. Root is the common ancestor of all taxa. Sister taxa is a systematic term from cladistic denoting the closest relative of a group in phylogenetic tree. Species A and B joining together represents the closely related relationship between them. The star marked represents a recent common ancestor of species A and B. A star marked is more recently related than two ones. Species F is the most distantly related of the taxa so called the "outgroup". The outgroup species are necessary to indicate the last common ancestor in the cladistic method.

corresponding to the most common ancestor of all of the entities at the terminal nodes or leaf nodes. This kind of the tree has an uncontroversial outgroup that is a monophyletic group of organisms closing enough to serve as a reference group for determination of the evolutionary relationship among three or more monophyletic groups of organisms. On the other hand, the unrooted tree illustrates only relatedness of the leaf nodes, without making any assumptions about common ancestry.[22,23] In the case of the absence of an outgroup, the most probable place for the root is the middle of the tree (midpoint) (Fig. 6.2).

There must be an understanding of the function of proteins and genes before starting the phylogenetic tree construction. Phylogeny is the evolution of a genetically related group of organisms via the study of protein or gene evolution by involving the comparison of homologous sequences. Homologs can be orthologs or paralogs. Orthologs are homologs produced by speciation, whereas paralogs are the ones produced by gene duplication (multigene families). If paralogs are used to infer species relationships, a misunderstanding of the phylogenetic relationships may occur because of the lack of some of the copies in duplicated genes.[21] To build phylogenetic trees, ortholog sequences need to be used for estimating the amount of evolution between these sequences.

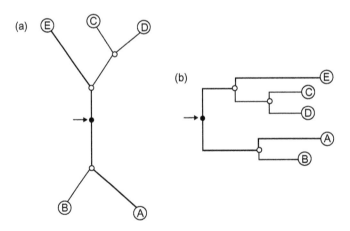

Figure 6.2 Midpoint rooting phylogenetic tree. The tree attempts to root the tree in its middle point. (A) Between A to E is the longest tip-to-tip distance. The root is then placed exactly half way between those two tips. (B) The topology redrawn from Fig. 6.2A. The point of rooting represents the ancestral point. This kind of tree is suitable for estimating phylogenetic relationship where the actual root is not known. All taxa in the tree were assumed to have constant "clock-like" rate of evolution.

6.3 HOW TO CONSTRUCT PHYLOGENETIC TREES

Steps for constructing a phylogenetic tree include: 1) assembling a dataset, 2) multiple sequence alignment, 3) determining the substitution model, 4) tree building, and 5) tree evaluation.

6.3.1 Assembling a Dataset

In phylogenetic analysis, public data should be retrieved in addition to your own sequence dataset that has to be prepared. For assembling a dataset from the public domain, nucleotide databases are mainly stored independently in the International Sequence Database Collaboration (INSDC), which includes the USA (GenBank), EU (ENA), and Japan (DDBL). Moreover, some of the most exciting molecular evolutionary data have been coming from genome sequencing projects (Table 6.3). To find a set of related sequences, there are two basic kinds of searching via search engines: usingkeywords and using similarity. A keyword search is the way to find identified sequences by looking through the written description in the annotation section of database files. Similarity is another way of searching; it involves looking directly at the sequences.[21] Software available for both search strategies is in Table 6.3.

6.3.2 Multiple Sequence Alignment

This is an important step for phylogenetic tree construction. The multiple sequence alignments are usually constructed by progressive sequence alignment.[24,25] This method builds the alignment by starting with the most similar sequences and adding up the more dissimilar divergent ones in a so-called "stepwise alignment." The alignment performs according to an explicitly phylogenetic criterion (a guide tree). The guide tree is included as a part of alignment, but it only shows how the alignment was assembled.[21]

Recently, many programs have become available, but the "Clustal" program is well-known as the easiest and the most widely used. An alternative software is the "MUSCLE" program, which is significantly faster than Clustal, especially for larger alignments. The alignment file from Clustal can be imported into BioEdit for sequence editing.[26,27]

6.3.3 Determination of Substitution Model

A phylogenetic tree is a kind of molecular archaeology that tries to reconstruct possible evolutionary relationships by extrapolating backward from a small dataset from surviving organisms. The problem is

Table 6.3 Bioinformatic Resources for Molecular Systematics

Databases
Primary nucleotide databases
DDBJ (Japan): http://www.ddbj.nig.ac.jp/
ENA (EU): www.ebi.ac.uk/ena/
GenBank (USA): http://www.ncbi.nlm.nih.gov/
Genomes
NCBI: http://ncbi.nlm.nih.gov/genome
Bioinformatic Harvester: http://harvester.kit.edu
Ensembl: www.wnsembl.org
Reference sequence databases (for non-redundant and well-annotated set of sequences)
RefSeq: www.ncbi.nlm.nih.gov/refseq
Data acquisition (search engines)
SRS@EMBL-EBI: http://srs.ebi.ac.uk
Entrez: http://www.ncbi.nlm.nih.gov/Entrez/
BLAST: http://www.ncbi.nlm.nih.gov.BLAST/
Multiple sequence alignment
ClustalX: www.clustal.org
MUSCLE: www.drive5.com/muscle/
T-coffee: www.tcoffee.org
DNA sequence editor
BioEdit: http://www.mbio.ncsu.edu/bioedit/bioedit.html
DNAAlignEditor: http://maize.agron.missouri.edu/ ~ hsanchez/DNAAlignment_Tool.html
Best-fit model selection
jModelTest2: http://code.google.com/p/jmodeltest2/
MEGA: www.megasoftware.net
Phylogenetic analysis
PAUP: http://paup.csit.fsu.edu/index.html
PHYLIP: http://evolution.genetics.washington.edu/phylip.html
MEGA: www.megasoftware.net
MrBayes (for Bayesian inference of phylogeny) : mrbayes.sorceforge.net/index.php
Note: Programs for molecular phylogeny have been updated and listed in http://evolution.genetics.washington. edu/phylip/software.html

the true evolutionary differences between two sequences because of multiple mutations, especially at the more rapidly evolving sites. Therefore, the various models of nucleotide substitutions have been developed to estimate the biggest possible difference between sequences based on the current data.[21] Programs available for selecting best-fit models of nucleotide substitution are listed in Table 6.3.

6.3.4 Tree Building

There are two general categories of methods for constructing phylogenetic trees: clustering methods and tree searching methods. The clustering methods are known as the distance-matrix method, in which the UPGMA and neighbor-joining methods are generally used. The distance-matrix method requires the genetic distance, which is determined for all pairwise combinations of OTUs and then those distances are assembled into a tree. The tree searching methods are known as discrete data methods. Maximum parsimony, maximum likelihood, and Bayesian inference methods are applied directly to nucleotide sequences. Discrete data methods examine the nucleotide variation in each column of the alignment separately and consider only "Phylogenetically informative sites" for searching the best tree that conforms to all of the information. Based on the algorithm differences, distance-matrix methods are much faster than tree searching methods. The clustering methods, however, only presume the most closely related among organisms, whereas discrete data analyses try to find a set of all possible classification schemes and then measure how the characteristics evolve on each of all possible trees.[20,21,23] The user friendly programs for constructing phylogenetic trees are listed in Table 6.3.

The most reliable tree is required for revealing the most probable evolutionary relationships among organisms. Therefore, the trees are always constructed by more than one phylogenetic method and then the congruent phylogenetic relationships found can be supported with high confidence.

6.3.5 Tree Evaluation

The simplest test of phylogenetic accuracy is the bootstrapping method. This method is a statistical technique developed to support each of the relationships in the tree. In bootstrapping, the original data matrix is randomly resampled with replacements to produce pseudo-replicate datasets. The tree-building algorithm is performed on each of these replicate datasets. If the bootstrapping offers a measurement of which parts of the tree are weakly supported, a grouping of such relationships presents in a low percent of the bootstrap replicates. This implies that if another dataset were collected, there is a good chance that the group would not be recovered. Bootstrapping can be used to assess the strength of support in virtually any type of analysis. Competitive results depending on groups of relationships with low bootstrapping support should be viewed with caution. Normally, the higher bootstrapping proportion is clearly better for interpreting the relationship, but what is a reasonable cut-off? In the case when bootstrapping proportions are conservative measures of support; a 70% value of bootstrapping might indicate strong support for a group of relationships.[28]

The bootstrapping proportions help to predict whether the same result would be seen if more data were collected. Although high bootstrapping proportions are a necessity, they are not sufficient. So, bootstrapping cannot be used to overcome an inappropriate analysis of the data. Figure 6.3 shows that the bootstrapping approach involves the generation of pseudo-replicate datasets by resampling by replacing the sites in the original data matrix.[29] Bootstrapping represents the value of interpreting the number of cases in which the sequences are classified together. The values should be displayed as percentages and only values of 50% and higher, because of this will lead to easier understanding and comparison with other trees.[30]

6.4 FURTHER PERSPECTIVES

Since the ortholog region is the most suitable molecular marker for underlying speciation events but not for gene duplication events, orthology prediction is an important part of phylogenomic approaches.[31] One thing that should be a primary concern is that the genes selected as genetic markers must be found in all taxa under study. Another important issue is the genes selected should have constant state characteristic frequencies and substitution rates in all studied lineage over time. The two criteria mentioned above are required for systematics and population genetic studies, but are very difficult to discover in nuclear genomes.

At the time that this chapter was written, many complete or partially complete mitochondrial genomes have been published. By comparative analyses of complete mitochondrial genomes of metazoans, it was revealed that genes/regions of the genomes have the potential to be good genetic markers. This is because gene duplications in mitochondrial genomes occur rarely. Therefore, orthology prediction is

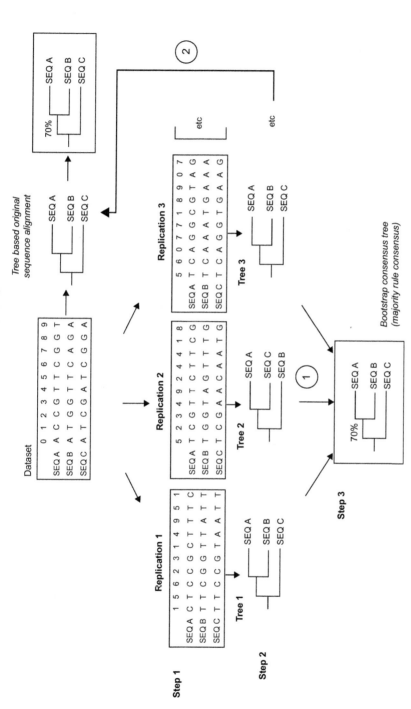

Figure 6.3 Process of bootstrap analysis. There are 3 steps starting from the original sequence alignment dataset. Step 1: the dataset is randomly sampled with replacement to crete multiple pseudo-datasets of the same size as the original so called "replicate". Pseudo-datasets are assembled by numbers of datasets created. Step 2: Individual trees are constructed from each pseudo-dataset. Step 3: There are 2 options for summarizing the results of bootstrapping. The option 1: each of the pseudo-dataset tree are scored by numbers of each node appearing (grouping). The bootstrap consensus tree are summarized as a majority-rule consensus tree. Frequency of node grouping represents as bootstrap support (percentage) on each node. The option 2: All bootstrap tree are compared with the tree constructed based on the original alignment and the number of times forming a cluster (as defined in the original tree) is then superimposed on the original tree as bootstrap support.

apparently an easy task, especially for complete mitochondrial genomes.[32] The mitochondrial genomes are comprised of comparative high mutation rates and mixtures of conserved and variable sites that facilitate the application of universal primer sets, providing sufficient phylogenetic signals.[33] Moreover, the lack of recombination and maternal inheritance make mitochondrial genes suitable molecular markers for inferring population structure.[34,35]

The mitochondrial genomes of metazoans are available in the NCBI Reference Sequence Database (NCBI RefSeq). This database is a collection of genomic, transcript, and protein sequence records that are selected from public sequence archives.[36] Up-to-date information on mitochondrial genomes of some important trematodes have been completed, such as mitochondrial genomes of *Schistosoma haematobium*, *S. mansoni*, *S. japonicum*, *Clonorchis sinensis*, *Opisthorchis viverrini*, *O. felineus*, *Fasciolopsis buski*, and *Fasciola hepatica*. The whole genomes completed above can find conserved mtDNA sequences for future investigation of taxonomy, systematics, and also population genetics of fish-borne trematodes.

REFERENCES

1. Tautz D, Arctander P, Minelli A, Thomas RH, Vogler AP. A plea for DNA taxonomy. *Trends Ecol Evol* 2003;**18**:70−4.

2. Hwang U-W, Kim W. General properties and phylogenetic utilities of nuclear ribosomal DNA and mitochondrial DNA commonly used in molecular systematics. *Korean J Parasitol* 1999;**37**:215−28.

3. Le TH, Blair D, Donald P, et al. Mitochondrial genomes of human helminths and their use as markers in population genetics and phylogeny. *Acta Trop* 2000;**77**:243−56.

4. Olsen PD, Cribb TH, Tkach VV, et al. Phylogeny and classification of Digenean (Plathyhelminthes: Trematoda). *Int J Parasitol* 2003;**33**:733−55.

5. Nolan J, Cribb TH. The use and implications of ribosomal DNA sequencing for the discrimination of Digenean species. *Adv Parasitol* 2005;**60**:101−63.

6. Sunnuck P. Efficient genetic markers. *Trends Ecol Evol* 2000;**15**:199−203.

7. Templeton RCA, Constantine CC, Morgan UM. Overview and significance of molecular methods: what role for molecular epidemiology? *Parasitol* 1998;**117**:161−75.

8. Constatine CC. Importance and pitfalls of molecular analysis to parasite epidemiology. *Trends Parasitol* 2003;**19**:346−8.

9. Blair D. Use of molecular data in Trematode systematics: strengths and weaknesses. *Acta Parasitol Turcica* 1996;**20**:375−86.

10. Qu LH, Nicoloso M, Bachellerie JP. Phylogenetic calibration of the 5' terminal domain of large rRNA achieved by determining twenty eukaryotic sequences. *J Mol Evol* 1988;**28**: 113–24.

11. Barker SC, Blair D, Garrett AR, et al. Utility of the D1 domain of nuclear 28 S rRNA for phylogenetic inference in the Digenea. *Syst Parasitol* 1993;**26**:181–3.

12. Shylla JA, Ghatani S, Tandon V. Utility of divergent domains of 28 S ribosomal RNA in species discrimination of paramphistomes (Trematoda: Digenea: Paramphistomoidea). *Parasitol Res* 2013;**112**:4239–53.

13. Morgan JAT, Blair D. Nuclear rDNA ITS sequence variation in the trematode genus Echinostoma: an aid to establishing relationships within the 37-collar-spine group. *Parasitol* 1995;**111**:609–15.

14. Avise JC. *Introduction. Molecular markers, natural history and evolution.* Chapman and Hall; 1994. pp. 3–15.

15. Boore JL. Animal mitochondrial genome. *Nucl Acids Res* 1999;**27**:1767–80.

16. Shadel GS, Clayton DA, Mitochondrial DNA. maintenance in vertebrates. *Ann Rev Biochem* 1997;**66**:409–35.

17. Saijuntha W, Sithithaworn P, Wongkham S, et al. Mitochondrial DNA sequence variation among geographical isolates of Opisthorchis viverrini in Thailand and Lao PDR, and phylogenetic relationships with other trematodes. *Parasitol* 2008;**135**:1479–88.

18. Thaenkham U, Nuamtanong S, Sa-nguankiat S, et al. Monophyly of Opisthorchis viverrini populations in the lower Mekong Basin, using mitochondrial DNA nad1 gene as the marker. *Parasitol Int* 2010;**59**:242–7.

19. Sithithaworn P, Andrews RH, Petney TN. The systematics and population genetics of Opisthorchis viverrini sensu lato: implications in parasite epidemiology and bile duct cancer. *Parasitol Int* 2012;**61**:32–7.

20. Brinkman FS, Leipe DD. *Phylogenetic analysis: in Bioinformatics: a practical guide to the analysis of genes and proteins.* 2nd ed. New York: John Wiley and Sons, Inc.; 2001.

21. Baldauf SL. Phylogeny for the faint of heart: a tutorial. *Trends Genet* 2003;**19**:345–51.

22. Gregory TR. Understanding evolutionary trees. *Evo Edu Outreach* 2008;**1**:121–37.

23. Page RDM, Holmes EC. *Molecular evolution: phylogenetic approach.* Oxford: Blackwell Science Ltd.; 1998.

24. Feng DF, Doolittle RF. Progressive sequence alignment as a prerequisite to correct phylogenetic trees. *J Mol Evol* 1997;**25**:351–60.

25. Higgins DG, Sharp PM. CLUSTAL: a package for performing multiple sequence alignment on a microcomputer. *Gene* 1988;**73**:237–44.

26. Higgins D, Thompson J, Gibson T, et al. CLUSTAL W: improving the sensitivity of progressive multiple sequence alignment through sequence weighting, position-specific gap penalties and weight matrix choice. *Nucl Acids Res* 1994 Nov;**22**:4673–80.

27. Edgar RC. MUSCLE: a multiple sequence alignment method with reduced time and space complexity. *BMC Bioinf* 2004;**5**:113.

28. Felsenstein J. Confidence limits on phylogenies: an approach using the bootstrap. *Evolution* 1985;**39**:783–91.

29. Holder M, Lewis PO. Phylogeny estimation: traditional and Bayesian approaches. *Nat Rev* 2003;**4**:275–84.

30. Michu E. A short guide to phylogeny reconstruction. *Plants Soil Environ* 2007;**53**:442−6.

31. Altenhoff AC, Dessimoz C. Inferring orthology and paralogy. In: Anisimova M, editor. *Evolutionary genomics: statistical and computational methods, methods in molecular biology.* Springer Humana; 2012. p. 855.

32. Bernt M, Bleidornb C, Braband A. A comprehensive analysis of bilaterian mitochondrial genomes and phylogeny. *Mol Phylo Evol* 2013. Available from: http://dx.doi.org/10.1016/j.ympev.2013.05.002.

33. Moritz C, Dowling TE, Brown WM. Evolution of animal mitochondrial DNA: relevance for population biology and systematics. *Ann Rev Eco Syst* 1987;**18**:269−92.

34. Biswall DK, Ghatani S, Shylla JA. An integrated pipeline for next generation sequencing and annotation of the complete mitochondrial genome of the giant intestinal fluke, Fasciolopsis buski (Lankester, 1857) Looss, 1899. *Peer J* 2013. Available from: http://dx.doi.org/10.7717/peerj.207.

35. Sultana T, Kim J, Lee SH. Comparative analysis of complete mitochondrial genome sequences confirms independent origins of plant-parasitic nematodes. *BMC Evol Biol* 2013;**13**:12.

36. Pruitt KD, Tatusova T, Brown GR, et al. NCBI Reference Sequences (RefSeq): current status, new features and genome annotation policy. *Nucl Acids Res* 2012;**40**:D130−5.

Table 1 Reported Hosts of Small Liver Flukes

Liver Fluke	Host
Amphimerus sp.[1]	**Final host**
	Cat, dog, opossum (*Phillander Opossum*), rodent (*Nectomys squamipes*)(Cricetidae)
Clonorchis sinensis[2]	**1st intermediate host**
	Parafossarulus manchouricus, Alocinma longicornis, Bithynia fuchsiana
	2nd intermediate host
	Abbottina spp., *Acanthorhodeus gracilis, Acheilognathus taenianalis, Acheilognathus lanceolata, Carassius* spp., *Ctenopharyngodon idellus, Cyprinus carpio, Gnathopogon* spp., *Hemibarbus labeo, Hemiculter* spp., *Mylopharyngodon aethiops, Pungtungia herzi, Pseudogobio esocinus, Pseudorasbora parva, Gnathopogon elongatus, Rhodeus* spp., *Sarcocheilichthys sinensis, Sarcocheilichthys variegatus*
	Final host
	Pig, dog, cat, civet, hare, rodent
Metorchis bilis[3]	**1st intermediate host**
	Bithynia tentaculata, B. inflata, Bithynia troschelii
	2nd intermediate host
	Ide, roach, dace, tench, minnow, gudgeon, verkhovka, silver crucian carp
	Final host
	white tailed eagle, red fox, American mink, American muskrat, otter
Metorchis conjunctus[4,5]	**1st intermediate host**
	Amnicola limosa
	2nd intermediate host
	Catostomus catostomus, Catostomus commersoni, Perca flavescens, Salvelinus fontinalis, Semotilus corporalis
	Final host
	Raccoon, gray fox, mink, wolf, dog
Metorchis orientalis[6]	**1st intermediate host**

	2nd intermediate host
	Pseudorasbora parva
	Final host
	duck, cat dog

(*Continued*)

Table 1 (Continued)

Liver Fluke	Host
Opisthorchis felineous[2,3]	**1st intermediate host**
	Bithynia leachii, Bithynia inflata, Bithynia troschelii
	2nd intermediate host
	Abramis bramae, A. sapa, Aspius aspius, Barbus barbus, Leuciscus idus, Leuciscus leuciscus, Phoxinus spp., *Rutilus rutilus, Tinca tinca*, 23 species of carp, including: ide, dace, roach
	Final host
	Cat, fox, muskrat, corsac fox, dog (28 species of mammals)
	Chipmink, beaver, Caspian seal, wild pig, domestic pig, and lab. Hedgehog, rabbit, guinea-pig, house mouse, golden hamster, black-bellied hamster.
Opisthorchis viverrini[2,7,8]	**1st intermediate host**
	Bithynia funiculata, B. goniophalos, Bithynia siamensis
	2nd intermediate host
	Barbonymus altus, Barbonymus brevis, Barbus leiacanthus, Cyclocheilichthys apogon, Cyclocheilichthys armatus, Cyclocheilichthys enoplos, Cyclocheilichthys repasson, Esomus metallicus, Hampala dispar, Hampala macrolepidota, Henicorhynchus siamensis, Labiobarbus siamensis, Puntioplites proctozysron, Systomus orphoides, Thynnichthys thynnoides
	Final host
	Cat, dog

Table 2 Hosts of fish-borne Echinostomes

Parasite	Hosts
Echinochasmus fujianensis	**1st intermediate host**
	Bellamya aeruginosa[9]
	2nd intermediate host
	Pseudorasbora parva, Cyprinus carpio[9]
	Final host
	Dog, cat, pig, rat[10]
Echinochasmus japonicus	**1st intermediate host**
	Parafossarulus manchouricus[11]
	2nd intermediate host
	Gnathopogon elongatus, Misgurnus anguillicaudatus, Odontobutis obscura, Zacco platypus, Pseudorasbora parva[12]
	Final host
	Dog, cat, rat, mouse, duck, egret[13]

(Continued)

Table 2 (Continued)	
Parasite	**Hosts**
Echinochasmus liliputanus	**1st intermediate host**
	Parafossarulus striatulus[9]
	2nd intermediate host
	Pseudorasbora parva[9]
	Final host
	Badger, fox, dog, cat[9]
Echinochasmus perfoliatus	**1st intermediate host**
	Parafossarulus manchouricus, Bithynia leachii, Lymnaea stagnalis[14]
	2nd intermediate host
	Gnathopogon elongatus, Misgurnus anguillicaudatus, Odontobutis obscura, Pseudorasbora parva, Zacco platypus[12]
	Final host
	Fox, dog, cat, rat, fowl, wild boar[15,16]
Echinostoma cinetorchis	**1st intermediate host**
	Cipangopaludina chinensis malleata, Hippeutis cantori,[17] *Polypylis hemisphaerula,*[w] *Segmentina hemisphaerula*
	2nd intermediate host
	Misgurnus anguillicaudatus[2]
	Final host
	Dog, rat[18]
Echinostoma hortense	**1st intermediate host**
	Austropeplea ollula,[w] *Lymnaea previa,*[19] Radix sp.[20]
	2nd intermediate host
	Misgurnus spp.[19]
	Final host
	Dog, cat, rat[19]

Table 3 List of Hosts of Heterophyid Intestinal Flukes Infecting Humans	
Parasite	**Host**
Apophallus donicus	**1st intermediate host**
	Fluminicola virens[21]
	2nd intermediate host
	Acerina cernua,[22] *Catostomus macrocheilus,*[21] *Gobio fluviatilis,*[22] *Sander lucioperca,*[22] *Oncorhynchus kisutch,*[21] *Perca fluviatilis,*[22] *Ptychocheilus oregonensis,*[21] *Richardsonius balteatus,*[21] *Rhinichthys osculus nubilus,*[21] *Salmo gairderi,*[21] *Scardinius erythrophthalmus*[21]
	Final host
	Alopex lagopus,[23] *Gallus gallus domesticus,*[21] *Mustela sarmatica,*[22] *Rattus norvegicus,*[24] *Vulpes lagopus*[25]
	Rabbit,[21] white mouse,[21] gerbil,[21] golden hamster,[21] cat,[25] dog[25]

(Continued)

Table 3 (Continued)

Parasite	Host
Ascocotyle longa	**1st intermediate host**
	Heleobia australis[26]
	2nd intermediate host
	Mugil cephalus,[27] *Mugil curema,*[28] *Mugil incilis,*[26] *Mugil liza,*[26] *Mugil platanus,*[29] *Mugil trichodon*[28]
	Final host
	Casmerodius albus,[27] *Leucophoyx thula,*[27] *Lutra reponda,*[27] *Pelecanus occidentalis,*[27] *Phalacrocorax carbo,*[27] *Sula leucogaster,*[27] *Vulpes lagopus,*[30] cat,[27] chicken,[27] dog,[27] egret,[27] hamster,[26] mink,[27] night heron,[27] opossum,[27] white rat,[27] wolf[27]
Centrocestus armatus	**1st intermediate host**
	Semisulcospira japonica,[31] *Semisulcospira libertina, Semisulcospira multigranosa, Semisulcospira reiniana*
	2nd intermediate host
	Abbotina rivularis,[32] *Acheilognathus lanceolata, Acheilognathus tabila, Anabas testudineus, Aphyocypris chinensis, Carassius carassius, Channa argus, Cyprinus carpio, Gnathopogon elongates, Gobius similis, Leuciscus hakonensis, Misgurnus anguillicaudatus, Ophicephalus maculatus, Opsariichthys uncirostris, Paracheilognathus rhombea, Pelteobagrus fulvidraco, Pseudoperilampus typus, Pseudorasbora parva, Rhodeus atremius, Rhodeus ocellatus, Rhodeus oryzae, Tribolodon hakonensis, Zacco platypus, Zacco temminckii*
	Final host
	Ardea cinerea,[33] *Egretta intermedia, Milvus migrans lineatus, Nycticorax nycticorax,*[34] *Phalacrocorax carbo hanedae, Rattus norvegicus,*[35] cat,[14] dog, rabbit
Centrocestus cuspidatus	**1st intermediate host**
	Thiara sp.[36]
	2nd intermediate host
	Astatotilapia desfontainii,[36] *Gambusia* spp.[37]
	Final host
	Milvus aegyptius,[38] *Milvus parasitus,*[39] *Nycticorax nycticorax,*[36] dog, rat, guinea pig
Centrocestus formosanus	**1st intermediate host**
	Stenomelania newcombi,[37] *Semisulcospira libertina,*[40] *S. hidachiensis*
	2nd intermediate host
	Amblypharyngodon mola,[41] *Anabus testudineus,*[35] *Babeo bota, Carassius auratus, Carassisus carassius, C. tadiana, Cirrhina reba, Clarias fuscus, Ctenopharyngodon idella, Cyclocheilichthys* sp., *Cyprinus carpio, Gambusia affinis, Glossogobius giuris, Hampala dispar, Hemiramphus dussumieri, Kuhlia sandvicensis,*[37] *Labeo bota,*[35] *Limia caudofasciata, Macropodus opercularis, Misgurnus anguillicaudatus, Mugil cephalus,*[37] *Ophicephalus maculatus,*[42] *Ophicephalus striatus,*[35] *Ophicephalus tadianus, Parasilurus asotus, Polyacanthus operculatus, Pseudorasbora parva, Puntius semifasciolatus, Rhodeus ocellatus, Therapon plumbeus, Xiphophorus helleri,*[37] *Zucco platypus,*[35] *Bufo melanostictus, Rana limnocharis*
	Final host
	Ardea purpurea manilensis,[43] *Bubulcus ibis coromandus,*[35] *Egretta intermedia,*[34] *Rattus rattus,*[44] *Nyctereutes procyonoides,*[35] *Nycticorax nycticorax, Nyctereutes procyonoides,*[34] *Pyerreroides manillensis,*[35] guinea pig, pigeon, rabbit, chicken, cat, dog[42]

(Continued)

Table 3 (Continued)

Parasite	Host
Centrocestus longus	**1st intermediate host**
	Thiara sp.[45]
	2nd intermediate host
	Acanthogobius sp.,[46] *Anabas testudineus,*[45] *Rhinogobius* sp.,[46] *Clarias fuscus,*[45] *Liza* sp.,[46] *Misgurnus* sp.,[45] *Mugil* sp.,[46] *Ophicephalus maculatus,*[45] *Pleuronectes* sp.[46]
	Final host
	Cat,[46] dog[45]
Cryptocotyle lingua	**1st intermediate host**
	Littorina littorea,[47] *Littorina rudis, Littorina scutulata, Littorina sitkana,*[48] *Paludestina* sp.,[49] *Peringia ulvae*
	2nd intermediate host
	Acanthocottus aeneus,[49] *Cottus scorpius, Ctenolabrus adspersus, Gadus morhua, Gobius flavescens, G. minutus scorpius, Gadus morhua gallanis, Gobius ruthensparri,*[50] *Halichoerus grypus,*[51] *Labrus bergylta,*[50] *Lophopsetta maculata,*[49] *Macrozoarces americanus,*[50] *Menidia notala,*[49] *Menticirrhus saxatilis, Microgadus tomcod, Mullus auratus, Onos mustela,*[50] *Onos tricirratus, Osmerus mordax,*[49] *Phoca caspica,*[52] *Phoca vitulina, Pholis gunnellus,*[49] *Platichthys flesus, Pleuronectes platessa, Pollachius virens, Pomatomus saltatrix, Pronotus triacanthus, Pseudopleuronectes americanus, Scomber scombrus, Tautoga onitis, Tautogolabrus adspersus,*[47] *Trachinotus falcatus*[49]
	Final host
	Alca torda,[44] *Butorides virescens,*[49] *Columbus auritus, Gavia immer, Larus argentatus, Larus argentatus smithsonianus,*[34] *Leucophaeus atricilla,*[49] *L. canus, Larus delawarensis, Liasis fuscus, Litsea glaucescens, Larus marinus, Lutra lutra,*[53] *Lutreola vision,*[54] *Nycticorax nycticorax,*[49] *Podiceps auritus, Rissa tridactyla, Sterna dougallii, Sterna hirundo, Uria aalge, Vulpes fulva,*[55] *Vulpes lagopus,* cat,[49] dog
Haplorchis pumilio	**1st intermediate host**
	Semisulcospira libertina,[56] *Thiara tuberculata, Melania reiniana var. hitachiens*
	2nd intermediate host
	Acanthogobius flavimanus,[56] *Ambassis buruensis, Amphacanthus javus, Astatotilapia desfontainii,*[36] *Barbus canis, Barbus longiceps, Carassius auratus, Channa formosana, Clarias fuscus, Cyprinus carpio, Gambusia affinis, Gerres filamentosus,*[56] *Glossogobius giuris, Macropodus opercularis, Mugil affinis,*[36] *Mugil capito, Mugil cephalus, Ophicephalus striatus,*[56] *Parasilurus asotus,*[57] *Polyacanthus operculatus, Pseudorasbora parva, Puntius semifasciolatus,*[58] *Rhodeus ocellatus, Ophicephalus tadianus, Teuthis jarvus, Therapon plumbeus,*[56] *Tilapia galilaea,*[36] *Tilapia nilotica, T. simonsi, Zacco platypus*[57]
	Final host
	Cavia porcellus,[57] *Crocidura olivieri,*[29] cat, dog, *Hydromys chrysogaster,*[59] *Laras* sp.,[60] *Microcarbo melanoleucos,*[59] *Milvus migrans,*[57] *M. m. aegyptius, Milvus parasiticus, Mus musculus, Nycticorax nycticorax, Oryctolagus cuniculus, Pelecanus crispus, Pelecanus onocrotalus, Rattus norvegicus, Vulpes vulpes*[29]

(Continued)

Table 3 (Continued)

Parasite	Host
Haplorchis taichui	**1st intermediate host**
	Semisulcospira libertina, Tarebia granifera, Melania obliquegranosa, Melanoides tuberculata[2]
	2nd intermediate host
	Amblypharyngodon mola,[41] *Anabas testudineus,*[61] *Carassius auratus,*[62] *Channa formosana,*[32] *Cirrhina reba,*[41] *Ctenophryngodon idellus,*[32] *Cyclocheilichthys repasson, Cyprinus auratus, Cyprinus carpio, Gambusia affinis, Hampala dispar, Henicorhynchus siamensis, Labeo bata,*[41] *Labiobarbus leptocheila, Macropodus opercularis,*[61] *Misgurnus anguillicaudatus,*[32] *Mugil affinis, Mystacoleucus marginatus, Ophicephalus striatus, Oryzias latipes,*[61] *Parasilurus asotus,*[32] *Pseudorasbora parva, Puntius binotatus, Puntius brevis, Puntius gonionotus, Puntius leicanthus, Puntius orphoides, Puntius palata, Puntius semifasciolatus,*[58] *Raiamas guttatus, Rhodeus ocellatus,*[62] *Zacco platypus*
	Final host
	Felis viverrina,[63] *Milvus migrans,*[56] *M. m. aegyptius,*[38] *Pseudogyps bengalensis,*[64] rat,[41] pigeon, cattle egret, cat,[62] chicken, duckling, dog[42]
Haplorchis yokogawai	**1st intermediate host**
	Stenomelania newcombi,[65] *Melanoides tuberculata*[2]
	2nd intermediate host
	Ambassis buruensis,[66] *Amphacanthus javus,*[43] *Amblypharyngodon mola,*[41] *Arius manillensis,*[66] *Boleophthalmus pectinirostris,*[42] *Carassius sp., Cirrhina reba,*[41] *Clarias batrachus,*[66] *Clarias fuscus,*[42] *Cyprinus carpio, Cyclocheilichthys armatus,*[2] *Gerres kapas,*[66] *Hampala dispar,*[2] *Hemiramphus georgii,*[66] *Labeo bata,*[41] *Labiobarbus leptocheila,*[2] *Misgurnus sp.,*[42] *Mugil affinis, Mariaella dussumieri,*[62] *Mystus vittatus,*[67] *Onychostoma elongatum, Ophicephalus striatus,*[62] *Puntius ticto,*[67] *Therapon plumbeus*[66]
	Final host
	Ardea purpurea manilensis,[43] *Bubulcus ibis coromandus,*[62] *Corvus macrorhynchos,*[67] *Corvus splendens, Felis viverrina,*[8] *Hydromys chrysogaster,*[59] *Macacus cynomolgus,*[68] *Mivus migrans,*[29] *Nycticorax nycticorax,*[65] *Pithecus philippinensis,*[43] *Pyrreroidios manilensis,*[66] *dog*[63]
Heterophyes heterophyes	**1st intermediate host**
	Cerithidea cingulata microptera,[69] *Pirenella conica*[68]
	2nd intermediate host
	Barbus canis,[36] *Epinephelus aeneus, Gambusia affinis,*[68] *Lichia amia,*[60] *L. glauca, Mugil auratus, Mugil capito, Mugil cephalus, M. japonicus, Sciaena aquilla,*[70] *Solea vulgaris, Tilapia nilotica,*[68] *Tristramella simonis*[60]
	Final host
	Canis aureus lupaster,[70] *Circaetus gallicus,*[60] *Felis chaus nilotica,*[70] *Genetta tigrina,*[71] *Milvus aegyptius,*[60] *Pelecanus onocrotalus, Rattus norvegicus,*[72] *Rattus rattus,*[44] *Rhinolophus sp.,*[73] *Sterna hirundo, Vulpes vulpes, Zalophus californianus,*[74] *rabbit,*[38] dog, cat
Heterophyes kutsuradai	**1st intermediate host**

	2nd intermediate host
	Mugil cephalus,[75] *Zacco platypus*
	Final host

Table 3 (Continued)

Parasite	Host
Heterophyes nocens	**1st intermediate host**
	Tympanotonus microptera[62]
	2nd intermediate host
	Acanthogobius flavimanus,[32] *Glossogobius giuris brunneus, Liza menada, Mugil cephalus, M. japonicus, Therapon oxyrhynchus, Tridentiger obscurus*
	Final host
	Hydromys chrysogaster,[76] *Rattus norvegicus*,[77] dog, cat
Heterophyopsis continua	**1st intermediate host**
	Tympanotonus microptera[77]
	2nd intermediate host
	Acanthogobius flavimanus,[2] *Ambassis buruensis*,[62] *Amphacanthus javus, Anabas testudineus*,[42] *Atherina balabacensis*,[62] *Boleophthalmus pectinirostris*,[42] *Clupanodon punctatus, Coilia* sp.,[32] *Conger myriaster, Cyprinus carpio*,[42] *Dorosoma thrissa*,[32] *Gerres filamentosus*,[62] *Gerres kapas, Harengula zunasi*,[32] *Hemiramphus georgii*,[62] *Lateolabrax japonicus, Mugil affinis*,[42] *M. dussumieri*,[62] *Pelates quadrilineatus*[43]
	Final host
	Dog,[42] *Fregata ariel ariel*,[62] *Larus argentatus vegae*[57]
Metagonimus minutus	**1st intermediate host**
	2nd intermediate host
	Acheilognathus lanceolatus intermedius,[57] *Mugil cephalus*[78]
	Final host
	Dog[57], cat, *Mus musculus, Rattus norvegicus*
Metagonimus miyatai[79]	**1st intermediate host**
	Semisulcospira globus, Semisulcospira libertina, Semisulcospira dolorosa
	2nd intermediate host
	Morocco steindachneri, Zacco platypus, Zacco temminckii, Tribolodon taczanowskii, Plecoglossus altivelis,
	Final host
	Dog, hamster, *Vulpes vulpes japonica, Nyctereutes procyonoides viverrinus, Milvus migran lineatus*
Metagonimus takahashii[79]	**1st intermediate host**
	Semisulcospira coreana, Koreanomelania nodifila
	2nd intermediate host
	Carrassius carassius, Cyprinus carpio, Tribolodon taczanowskii, Lateolabrax japonicas
	Final host
	Dog, *Milvus migran lineatus*

(Continued)

Table 3 (Continued)

Parasite	Host
Metagonimus yokogawai	**1st intermediate host**
	Katayama nosophora,[60] *Pyradus cingulatus, Semisulcospira bensoni,*[80] *S. libertina, Semisulcospira plicosa, Thiara amurensis,*[57] *T. ebenima,*[60] *Thiara extensa, Thiara gottschei, Tarebia granifera, Semisulcospira nodiperda quinaria*
	2nd intermediate host
	Acanthogobius flavimanus,[32] *Acheilognathus lanceolata,*[81] *A. limbatus,*[32] *A. morioka,*[82] *A. rhombea,*[32] *A. tabira, Abramis ballerus, A. brama,*[44] *Anguilla japonica,*[32] *Aspius aspius,*[83] *Biwia zezera, Blicca bjoerkna,*[44] *Carassius auratus,*[84] *C. aureus,*[60] *C. carassius,*[44] *Chaenogobius urotaenia,*[32] *Channa argus, Chondrostoma nasus, Cobitis biwae, Coregonus ussuriensis,*[85] *Coreopera kawamebari,*[32] *Cottus pollux, Ctenophoryngodon idellus,*[62] *Culter erythropterus,*[32] *Cyprinus carpio,*[62] *Eleotris potamophilis,*[86] *Esox lucius,*[44] *Gnathopogon biwae,*[32] *G. elongatus, Gobio gobio, Hemibarbus barbus,*[32] *H. labeo,*[j] *Hypophthalmichthys molitrix, Idus idus,*[83] *Ishikauiu steenackeri,*[32] *Lateolabrax japonicas, Leuciscus borysthenicus, L. hakonensis,*[60] *L. idus,*[32] *L. waleckii,*[85] *Leiocassis brazhnikovi, L. ussuriensis, Lucioperca lucioperca, Morocco steindachneri,*[32] *Mesocottus haitjei,*[85] *Mesopus olidus, Misgurnus anguillicaudatus,*[d] *Mugil cephalus, Odontobutis obscura,*[82] *Oncorhynchus masou,*[32] *O. ehodurus, Ophicephalus argus, Opsariichthys uncirostris, Oryzias latipes, Parabramis pekinensis, Parasilurus asotus, Pelecus cultratus, Plecoglossus altivelis, Pseudopus leptocephalus,*[85] *Pseudobagrus fulvidraco, Pseudogobio esocinus, Pseudoperilampus typus, Pseudorasbora parva, Pungtungia herzi, Pyradus cingulatus, Rhodeus ocellatus,*[32] *Rhodeus oryzae, Rhodeus sinensis,*[86] *Rutilus rutilus,*[32] *Rutilis heckelii, Salangichthys microdon,*[81] *Salmo milktschish, Sarcocheilichthys variegatus, Scardinius erythrophthalmus,*[44] *Siniperca chuatsi, Staraia zbourievka,*[32] *Tribolodon hakonensis, Tridentiger obscurus, Vimba vimba, Xenocypris macrolepis, Zacco platypus, Z. temminckii*
	Final host
	Archibuteo lagopus,[57] *Ardea cinerea, A. purpurea, Buteo buteo, Cercopithecus* sp.,[87] *Chimarrogale platycephala,*[88] *Ciconia ciconia,*[57] *Egretta i. intermedia,*[34] *Felis pardus villosa,*[89] cat, dog,[44] *Haliaeetus albicilla,*[57] *Larus canus,*[90] *Larus ridibundus,*[91] *Milvus korschun,*[57] *M. migrans lineatus,*[34] *Mus molossinus,*[60] *M. musculus,*[44] *Nyctereutes procyonoides,*[57] *Nycticorax nycticorax,*[34] *Oreocincla dauma aurea,*[90] *Pelecanus crispus,*[92] *Phalacrocorax carbo,*[34] *P.c. hanedae, Podiceps ruficollis japonicas,*[57] *Puffinus leucomelas, P. nativitatis, Rattus norvegicus,*[88] *R. rattus*
Procerovum calderoni	**1st intermediate host**
	Thiara tuberculata chinensis,[93] *Thiara riquetti*
	2nd intermediate host
	Ambassis buruensis,[94] *Anabas testudineus, Creisson janthinopterus, Creisson validus, Eleutheronema tetradactylum, Gerres filamentosus, Glossogobius giurus, Hestia balabacensis, Hemiramphus georgii, Mollienesia latipinna, Mugil dussumieri, Ophicephalus striatus*
	Final host
	Dog,[95] cat
Procerovum varium[96]	**1st intermediate host**
	Thiara tuberculata
	2nd intermediate host
	Acanthogobius flavimanus, Hemiramphus sajori, Liza menada, Mugil cephalus, Oryzius melastigma
	Final host
	Cat, chick, duckling, mice, *Ardeola grayii*

(Continued)

Table 3 (Continued)

Parasite	Host
Pygidiopsis summa	**1st intermediate host**
	Tympanotonus microptera[97]
	2nd intermediate host
	Liza haematochila,[65] *L. menada,*[98] *Mugil cephalus, Pseudorasbora parva*[32]
	Final host
	Larus sp.,[94] *Macacus irus,*[82] *Macacus rhesus, Milvus migrans lineatus,*[29] *Nycticorax nycticorax,*[59] dog,[94] cat
Stellantchasmus falcatus	**1st intermediate host**
	Stenomelania newcombi,[32] *Tarebia granifera, T. g. mauiensis*[99]
	2nd intermediate host
	Acanthogobius flavimanus,[100] *Liza menada,*[100]
	Final host
	Crocidura olivieri,[29] *Hydromys chrysogaster,*[59] *Larus* sp.,[101] *Microcarbo melanoleuca,*[100] *Rattus norvegicus,*[59] dog,[100] cat
Stictodora fuscata	**1st intermediate host**

	2nd intermediate host
	Acanthogobius flavimanus,[32] *Mugil cephalus, Pseudorasbora parva*[99]
	Final host
	Cat,[77] dog[99]

Table 4 Reported Hosts of *Nanophyetus Salmincola*

Intestinal Fluke	Host
Nanophyetus salmincola	**1st intermediate host**
	Oxytrema silicula[102]
	2nd intermediate host
	Oncorhynchus kisutch[102,103]
	Oncorhynchus nerka[104]
	Salmo clarki clarki[104]
	Salmo clarki henshawi[104]
	Salmo clarki lewisi[104]
	Salmo gairdneri[104,105]
	Final host
	Dog, fox, coyote, raccoon, and golden hamster[102]

REFERENCES

1. Calvopina N, Cevallos W, Kumazawa H, et al. High prevalence of human liver infection by Amphimerus spp. Flukes, Ecuador. *Emerg Inf Dis* 2011;**17**:2331−4.

2. WHO. *Control of foodborne trematode infections. Reports of a WHO study Group.* Geneva: World Health Organization; 1995 (WHO Technical Report Series No 849).

3. Mordvinov VA, Yurlova NI, Ogorodova LM, et al. Opisthorchis felineus and Metorchis bilis are the main agents of liver fluke infection of humans in Russia. *Parasitol Int* 2012;**61**: 25−31.

4. MacLean JD, Arthur JR, Ward BJ, et al. Common-source outbreak of acute infection due to the North American liver fluke Metorchis conjuctus. *Lancet* 1996;**347**:154−8.

5. Wobesor G, Runge W, Stewart RR. Metorchis conjunctus (Cobbold, 1860) infection in wolves (Canis lupus), with pancreatic involvement in two animals. *J Wildlife Dis* 1983;**19**:353−6.

6. Lin J, Chen Y, Li Y, et al. The discovery of natural infection of human with Metorchis orientalis and the investigation of its focus. *Chin J Zoonoses* 2001;**17**:19−21.

7. Touch S, Komalamisra C, Radomyos P, et al. Discovery of Opisthorchis viverrini metacercariae in freshwater fish in southern Cambodia. *Acta Trop* 2009;**111**:108−13.

8. Aunpromma S, Tangkawattana P, Papirom P, et al. High prevalence of Opisthorchis viverrini infection in reservoir hosts in four districts of Khon Kaen Province, an opisthorchiasis endemic area of Thailand. *Parasitol Int* 2012;**61**:60−4.

9. Yu SH, Mott KE. Epidemiology and morbidity of food-borne intestinal trematode infections. *Trop Dis Bull* 1994;**91**:125−52.

10. Cheng YZ, Chen BJ, Fang YY, et al. A new species of Echinochasmus parasitic in man and observation of its experimental infection. *Wuyi Sci J* 1992;**9**:43−8, In Chinese.

11. Choi MH, Kim SH, Chung JH, et al. Morphological observations of Echinochasmus japonicas cercariae and the in vitro maintenance of its life cycle from cercariae to adults. *J Parasitol* 2006;**92**:236−41.

12. Komiya Y. Metacercariae in Japan and adjacent territories. In: Morishita K, Komiya Y, Matsubayashi H, editors. *Progress of medical parasitology in Japan*, vol. 2. Tokyo: Maruzen; 1965.

13. Sohn WM, Chai JY. Infection status with helminthes in feral cats purchased from a market in Busan, Republic of Korea. *Korean J Parasitol* 2005;**43**:93−100.

14. Yamaguti S. *Synopsis of digenetic trematodes of vertebrates*, vol. I. Tokyo: Keigaku Publishing; 1971.

15. Beaver PC, Jung RC, Cupp EW. *Clinical parasitology*. Philadelphia: Lea and Febiger; 1984.

16. Saeed I, Maddox-Hyttel C, Monard J, et al. Helminths of red fox (Valpes vulpes) in Denmark. *Vet Parasitol* 2006;**138**:168−79.

17. Chung PR, Jung Y, Park YK. Segmentina hemispherela: a new molluscan intermediate host for Echinostoma cinetorchis. *Korean J Parasitol* 1999;**87**:1169−71.

18. Cho SY, Kang SY, Ryang YS. Helminthes infection in the small intestine of stray dog in Ejungbu City, Kyunggi Do, Korea. *Korean J Parasitol* 1981;**19**:55−9, In Korean.

19. Lee SH, Noh TY, Sohn WM, et al. Chronological observation of intestinal lesions of rats experimentally infected with Echinostoma hortense. *Korean J Parasitol* 1990;**28**:45−52 In Korean.

20. Ahn YK, Kang HS. Development and cercarial shedding of Echinostoma hortense in the snail host, Radix auricularia coreana. *J Wonju Med Coll* 1988;**1**:137−52.

21. Niemi DR, Macy RW. The life cycle and infectivity to man of Apophallus donicus (Skrjabin and Lindtrop, 1919) (Trematoda: Heterophyidae) in Oregon. *Proc Helminth Soc Wash* 1974;**41**:223–9.

22. Modlinger G. Beitrage zur Biologie von Apophallus domicus. *Magy Biol Kutato Intezt Munkai* 1934;**7**:60–5.

23. Ericson AB. Parasites of beavers, with a note on synonym of Stichorchis subtriquetrus. *Am Midl Nat* 1944;**31**:625–30.

24. Harkema R. The parasites of some North Carolina rodents. *Ecol Monogr* 1936;**6**:153–232.

25. Price EW. Some new trematodes of the family Heterophyidae. *J Parasitol* 1932;**19**:166–7.

26. Simoes SBE, Barbosa HS, Santos CP. The life cycle of Ascocotyle (Phagicola) longa (Digenea: Heterophyidae), a causative agent of fish-borne trematodosis. *Acta Trop* 2010;**113**:226–33.

27. Scholz T. Taxonomis study of Ascocotyle (Phagicola) longa Ransom, 1920 (Digenea: Heterophyidae) and related taxa. *Syst Parasitol* 1999;**43**:147–58.

28. Conroy G, Perez KA. Report on the experimental infection of a smooth-headed capuchin monkey (Cebus apella) with metacercariae of Phagicola longa obtained from silver mullet. *Riv Ital Pisc Ittiopatol* 1985;**4**:154–5.

29. Kuntz RE, Chandler AC. Studies on Egyptian trematodes with special reference to the heterophyids of mammals. I. Adult flukes, with description of Phagicola longicollis n. sp., Cynodiplostomum namrui n. sp. and a Stephanoprora from cats. *J Parasitol* 1956;**42**:445–59.

30. Ransom BH. Synopsis of Heterophyidae with descriptions of a new genus and five new species. *Proc U S Nat Mus* 1920;**57**:527–73.

31. Ito J, Watanabe K. On the cercaria of Centrocestus armatus (Tanabe, 1922) Yamaguti, 1958, especially on its mucoid gland (Heterophyidae, Trematoda). *Jpn J Med Sci Biol* 1958;**11**:21–9.

32. Komiya Y, Suzuki N. The metacercariae of trematodes belonging to the family Heterophyidae from Japan and adjacent countries. *Kiseichugaku Zasshi* 1966;**15**:208–14 (In Japanese, English summary).

33. Yamaguti S. Studies on the helminth fauna of Japan. I. Trematodes of birds, reptiles and mammals. *Jpn J Zool* 1933;**5**:134.

34. Yamaguti S. Studies on the helminth fauna of Japan. Part 25. Trematodes of birds, IV. *Jpn J Zool* 1939;**8**:129–210.

35. Chen HT. The metacercaria and adult of Centrocestus formosanus (Nishigori, 1924), with notes on the natural infection of rats and cats with C. armatus (Tanabe, 1922). *J Parasitol* 1941;**28**:285–98.

36. Balozet L, Callot J. Trematodes de Tunisie. 3. Superfamilille Heterophoidea. *Arch Inst Pasteur Tunis* 1939;**28**:34–63.

37. Martin WE. Egyptian heterophyid trematodes. *Trans Am Micr Soc* 1959;**78**:172–81.

38. Gohar N. Les trematodes parasites du Milan egyptien Milvus migrans avec description d'une nouvelle espece et remarques sur les genres Haplorchis Looss, 1899 et Monorchotrema Nishigori, 1924. *Ann Parasitol Hum Comp* 1934;**12**:218–27.

39. Faust EC, Nishigori M. The life cycles of two new species of Heterophyidae, parasitic in mammals and birds. *J Parasitol* 1926;**13**:91–128.

40. Nishigori M. On a new trematode Stamnosoma formosanum n. sp. And its life history. *Taiwan Igakkai Zasshi* 1924;**234**:181–228, In Japanese.

41. Nath D, Pande BP. Identity of the three heterphyid metacercariae infesting some of the freshwater fishes (letter to editor). *Curr. Sci. Bangalore* 1970;**39**:325–6.

42. Kobayashi H. Investigations of trematodes in the South Sea Islands. II On trematodes of the genera Haplorchis and Centrocestus. *Taiwan Igakkai Zasshi, Taihoku* 1941;**40**:557—8.

43. Tubangui MA. A summary of the parasitic worms reported from the Philippines. *Philipp J Sci* 1947;**76**:225—322.

44. Bittner H, Sprehn CEW. Trematodes, Saungwurmer. *Biol Tiere Deutchl (Schulze) Lfg* 1928;**27**:133.

45. Kobayashi H. Studies on trematodes in Hainan Island. II Trematodes found in the intestinal tracts of dogs by experimental feeding with certain fresh and brackish water fish. *Jpn J Med Sci Path* 1942;**6**:187—227.

46. Onji Y, Nishio T. On intestinal distomes. *Iji Shimbun* 1916;**949**:589—93, In Japanese.

47. Cable RM. Studies on the germ-cell cycle of Cryptocotyle lingua. I. Germinal development in the larval stages. *Quart J Mier Sci* 1937;**76**:573—614.

48. Ching HL. Some digenetic trematodes of shore birds at Friday Harbour, Washington. *Proc Helminth Soc Wash* 1960;**27**:53—62.

49. Linton E. Notes on trematodes parasites of birds. *Proc US Nat Mus* 1928;**73**:1—36.

50. Rothchild M. A note on the life cycle of Cryptocotyle lingua (Creplin, 1825) (Trematoda). *Novitat Zool* 1939;**41**:178—80.

51. Duncan A. Notes on the food and parasites pf grey seals, Halichoerus gyrpus (Fabricyua), from the Isle of Man. *Proc Zool Soc Lond* 1956;**126**:635—44.

52. Shchupakov IG. Parasitic fauna of the Caspian seal. *Uhen Zapiski Leningr Gosudarstv Univ Bubnov Biol* 1936;**7**:134—43, [In Russian].

53. Fahmy MAM. On some helminth parasites of the otter, Lutra lutra. *J Helminthol* 1954;**28**:189—204.

54. McTaggart HS. Cryptocotyle lingua in British mink (letter to editor). *Naturec* 1958;**181**:651.

55. Swales WE. On Strptovitella acadiae (gen. et. Sp. Nov.) a trematode of the family Heterophyidae from the black duck (Anas rubripes). *J Helminthol* 1933;**11**:115—8.

56. Prakash R, Pande BP. On some of the known and hitherto unknown trematodes parasitic in the common pariah kite, Milvus migrans (Boddaert). *Indian J Helminthol* 1968;**20**:1—24.

57. Morozov FN. Superfamily Heterophyoidea Faust, 1929. In: Skryabin KI, editor. *Trematodes of animals and men*. Moscow: Akad Nauk Osnovy Trematodologii; 1952, In Russian.

58. Chen HT. Systematic consideration of some heterophyid trematodes in the subfamilies Haplorchinae and Stellantchasminae. *Ann Trop Med Parasitol* 1949;**43**:304—12.

59. Pearson JC. A revision of the subfamily Haplorchinae Looss, 1899 (Trematoda: Heterophyidae) I. The Haplorchis group. *Parasitol* 1964;**54**:601—76.

60. Witehberg G. Studies on the trematode family Heterophyidae. *Ann Trop Med Parasitol* 1929;**23**:131—239.

61. Hsu PK. A new trematode of the genus Procerovum from ducks and chickens in Canton (Trematoda: Heterophyidae). *Peking Nat Hist Bull* 1950;**19**:39—43.

62. Africa CM, de Leon W, Garcia EY. Complications in intestinal heterophidiasis in man. *Acta Med Philipp* 1940;**1**:132.

63. Kannangara DWW, Karunaratne GMS. A note on intestinal helminthes of dogs in Colombe. *Ceylon Vet J* 1970;**18**:47—9.

64. Gupta NK. On some digenetic trematodes of carnivorous birds in India. *Res Bull Punjab Univ Sci* 1967;**19**:305—14.

65. Martin WE. The life histories of some Hawaiian heterophyid trematodes. *J Parasitol* 1958;**44**:305–23.

66. Africa CM. Evidence of intramucosal invasion in the life cycle of Haplorchis yokogawai (Katsuta, 1932) Chen, 1936 (Heterophyidae). *J Philipp Isls Med Ass* 1937;**17**:734–7.

67. Pandet KC. Studies on metacercaria of freshwater fishes of India. I. on the morphology of metacercaria of Haplorchis yokogawai (Katsuta, 1932 Chen, 1936. *Proc Nat Acad Sci* 1966;**36**:437–40.

68. Khalil M. The effect of Heterphyese heterophyes on man. *Ann Rep Res Inst Endemic dis Hosp, Dept Pub Health* 1934;**3**:25–6.

69. Abbott RH. Hand book of medically important mollusks of the Orient and the Western Pacific. *Bull Mus Comp Zool Harvard Coll* 1948;**100**:245–328.

70. Wells WH, Randall BH. New hosts for trematodes of the genus Heterphyes in Egypt. *J Parasitol* 1956;**42**:287–92.

71. Tadros G. A collection of helminthes from mammals, birds and reptiles. *J Vet Sci* 1966;**3**: 101–10.

72. Fahmy MAM, Rifaat MA, Arafa MS. Results of a helminthic survey in a locality highly infected with rats near Cairo, UAR. *J Egypt Pub Health Ass* 1968;**43**:105–21.

73. Macy RW. First report of the human intestinal fluke Heterophyes heterophyes from a Yeman bat, Rhinolophus clivosus acrotis. *J Parasitol* 1953;**39**:517.

74. Ezzat MA, Tadros G. A contribution to the helminth fauna of Belgian Congo birds. *Ann K Mus Belg Kongo, Tervuren, S 8⁰ Zool Wetens* 1958;**69**:81.

75. Ozaki Y, Asada J. A new human trematode, Heterophyes katsuradai n. sp. *J Parasitol* 1925;**12**:216–8.

76. Waikagul J, Pearson JC. Heterophyes nocens Onji and Nishio, 1916 (Digenea: Heterophyidae) from the water rat, Hydromys chrysogaster (Geoff, 1804) in Australia. *Syst Parasitol* 1989;**13**:53–61.

77. Onji Y, Nishio T. On a new species of trematodes belonging to the genus Heterophyes. *Iji Shimbun* 1916;**954**:941–6, In Japanese.

78. Katsuta I. Studies on Formosan trematodes whose intermediate hosts are brackish water fishes. II. Metagonimus minutus n.sp., with mullet as its vector. *Taiwan Igakkai Zasshi* 1932;**31**:26–39, (In Japanese, English summary).

79. Shimazu T. Life cycle and morphology of Metagonimus miyatai (Digenea: Heterophyidae) from Nagano, Japan. *Parasitol Int* 2002;**51**:271–80.

80. Kagei N, Oshima T, Ishikawa K, Kihata M. Two cases of human infection with Stellantchasmus falcatus Onji & Nishio, 1915 (Heterophyidae) in Kochi Prefecture. *Jpn J Parasitol* 1964;**13**:472–8, In Japanese.

81. Okabe K. A synopsis of trematode cysts in freshwater fishes from Fukuoka Prefecture. *Fukuoka Ikwadaigaku Zasshi* 1940;**33**:309–35, In Japanese.

82. Takabayashi Y. Studies on trematode cysts in the fish in Yamaguchi Prefecture. *Nippon Kiseichu Gakkai Kiji* 1953;**21**:65–6, In Japanese.

83. Lane C, Low GC. *Trematodes infestations other than haemal. The practice of medicine in the tropics*, vol. 3. London: Byam W & Archibald RG; 1923.

84. Kobayashi H. On the intestinal parasites of the Korean, with special reference to some abnormal eggs. *Nippon No Ikai* 1920;**10**:889–93.

85. Zmeev GI. Les trematodes et les cestodes des poisson de l'Amour. *Parazit Sborn Zool Inst Akad Nauk* 1936;**6**:405–36.

86. Faust EC. Human intestinal parasites in north China. *Am J Hyg* 1929;**9**:505−8.

87. Mengert-Presser H. Over Metagonimus yokogawai inNed. Oost-Indie (Voorloopige mededeeling). *Herinneringsb Inst Trop Geneesk Leiden* 1924.

88. Dollfus RP. Distomiens parasites de Muridae de genre. *Mus Ann Parasitol* 1925;**3**:185−205.

89. Wu K. Helminthic fauna in vertebrates of the Hanchow area. *Peking Nat Hist Bull* 1937;**12**:1−8.

90. Morishita K. Some avain trematodes from Japan, especially from Formosa, with a reference list of all known Japanese species. *Ann Zool Jpn* 1929;**12**:143−73.

91. Shcherbovich IA. Trematodes of birds in the Far Eastern region. *Coll Pap On helm Skrj 40th year Sci edc And adm Achievem* 1946;296−300, In Russian.

92. Ianchev I. Untersuchungen uber eimige Helminthen und Helminthosen beiweissen Storchen und Pelikanen. *Izvest Zoo Inst Bulgr Akad Nauk Otdel Biol I Med Nauk* 1958;**7**:393−416 (Bulgarian text, Russian and German summaries).

93. Africa CM, Garcia EY. Heterophyid trematodes of man and dog in the Philippines with descriptions of three new species. *Philipp J Sci* 1935;**57**:253−67.

94. Vazquez-Colet A, Africa CM. Morphological studies on various Philippines heterophid metacercariae with notes on the incidence, site and degree of metacercarial infection in three species of marine fish. *Philipp J Sci* 1940;**72**:395−419.

95. Africa CM. Description of three trematodes of the genus Haplorchis (Heterophyidae), with notes on two other Philippines members of this genus. *Philipp J Sci* 1938;**66**:299−307.

96. Umadevi K, Madhavi R. Observations on the morphology and life-cycle of Procerovum varium (Onji & Nishio, 1916) (Trematoda: Heterophyidae). *Syst Parasitol* 2000;**46**:215−25.

97. Ochi G. Studies on the metacercaria of brackish water fishes. On the life history of Pygidiopsis summus. *Tokyo Iji Sinshi* 1931;**2712**:346−53, In Japanese.

98. Noda K. The larval development of Stellantchasmus falcatus (Trematoda: Heterphyidae) in the first intermediate host. *J Parasitol* 1959;**45**:635−42.

99. Hasegawa T. Uber die enzystierten Zerkaren in Paeudorasbora parva. *Okayama Igakki Zasshi* 1934;**46**:1397−434, (In Japanese, German summary).

100. Onj Y, Nishio T. On new intestinal trematodes. *Chiba Igakkai Zasshi* 1924;**2**:351−99, (In Japanese, German summary).

101. Miyazaki I. A survey of parasitic helminthes of house-dwelling rats in Kagoshima City. *Igaku To Seibutsugaku* 1946;**9**:219−20, In Japanese.

102. Bennington E, Pratt I. The life history of the salmon-poisoning fluke, Nanophyetus salmincola (Chapin). *J Parasitol* 1960;**46**:99−100.

103. Ferguson JA, et al. Survey of parasites in threatened stocks of coho salmon (Oncorhynchus kisutch) in Oregon by examination of wet tissues and histology. *J Parasitol* 2011;**97**:1085−98.

104. Baldwin NL, Millemann RE, Knapp SE. "Salmon poisoning" disease. III. Effect of experimental Nanophyetus salmincola infection on the fish host. *J Parasitol* 1967;**53**:556−64.

105. Eastburn RL, Fritsche TR, Terhune C. Human intestinal infection with Nanophyetus salmincola from salmonid fishes. *Am J Trop Med Hyg* 1987;**36**:586−91.

1 GENOMIC DNA PREPARATION

Genomic DNA extraction is the first important step for preparation of purified genomic DNA obtained from soft tissue. A nonorganic DNA extraction kit is recommended for purifying genomic DNA from a very small piece of tissue of adult worm and/ or metacercariae of a fish-borne trematode. The commercial nonorganic protocols avoid the toxicity inherent to phenol exposure in the process of traditional organic extraction. The researchers can extract genomic DNA following the manufacturers' instructions provided.

Therefore, this procedure herein instructs on sample preservation, DNA quantitation by spectrophotometer, and gel electrophoresis to determine DNA quality.

1.1 Preservation of Tissue Sample in Ethanol

It is recommended for the preservation of tissues of metacercariae/ adult worms, they are stored in 95–100% ethanol at -80 °C for long-term storage. The ethanol is not frozen, but this temperature is very cold for protecting DNA degradation. It is recommended that samples used in routine work are preserves at -20 °C. The metacercariae/ adult worms collected from a field can be preserved in 95–100% ethanol at room temperature (avoid warm/hot rooms). The container should be a small screw cap (including rubber o-ring) and should be sealed by paraffin tape.

1.2 DNA Quantitation by Spectrophotometer

The steps for determining the amount of DNA are:

Mix the DNA sample by gentle vortexing and inversion.
1) Add 5 µl of DNA sample to 495 µl of sterile water and mix well.
2) Place the diluted sample in a quartz microcuvette and measure the absorbance at 260 and 280 nm against a blank solution (DNA and protein absorb light maximally at 260 and 280 nm, respectively).

3) Compute the DNA concentration based on a formula: DNA concentration (μg/μl) = OD_{260} $\times$ 5.
4) The OD_{260} : OD_{280} ratio should be between 1.7 and 2.0. Lower values indicate protein contamination.

Note: For very tiny pieces of tissue, such as metacercariae or adult heterophyid intestinal flukes, DNA quantitation is not recommended.

1.3 Gel Electrophoresis to Determine DNA Quality

The DNA quality of purified genomic DNA can be determined before PCR amplification. This procedure is recommended for checking partially degraded DNA, especially in the case of high molecular weight DNA.

1) Prepare a 0.7% agarose gel in 1xTBE containing 0.5 μg/ml of ethidium bromide.
2) Mix an aliquot of the extracted DNA sample with a loading buffer and a loading dye and then load into a submerged well. Control samples representing intact and degraded DNA should be loaded into adjacent wells.
3) Electrophorese in 1xTBE buffer at 2 V/cm until the dye front reaches the end of the gel.
4) View the gel under UV light. Degraded DNA appears as a smear across the lane.

2 STANDARD POLYMERASE CHAIN REACTION (PCR) PROTOCOL

Herein, the optimized PCR condition is as follows:

2.1 Set up a 20–50 μl Reaction in a 0.2 ml Microfuge Tube with:

Reagent	Stock Solution	Final Concentration
ddH$_2$O	25 mM	1.5 mM
PCR buffer	10x	1x
dNTPs	10 mM each	0.2 mM each
Forward primer	100 μM	1 μM
Reverse primer	100 μM	1 μM
Taq polymerase	5U/μl	0.025U/μl
DNA template		10–20 ng/μl

2.2 Perform 25–35 cycles of PCR Using the Following Temperature Profile:

Initial denaturation	94°C; 5 min
Denaturation	94–95°C; 30 sec
Primer annealing	The choice of temperature is largely determined by the melting temperature (Tm) of the two PCR primers
Polymerase extension	72°C; The duration of this step is 30 sec for every 500 base in the PCR amplicon.
Final extension	72°C; 7–8°C

2.3 Validate the PCR Reactions

There are 2 ways for determining success or failure. The first is to simply take some of the final reaction and run it out on an agarose gel with an appropriate molecular weight marker. If the amplified product is the expected size relative to the marker, it is the right PCR target. The second is to directly sequence the PCR amplicon. This step is necessary in the case of studying an unknown species or confirming morphological identification.

Table 1. PCR Primers for Molecular Systematics and Molecular Identification of Fish-Borne Trematode

Parasite	DNA partition	Application	Primers (5′ → 3′)	References
Trematode universal 18 S	SSU (18 S)	MS	Fwd: Uni18S_F: GCTTGTCTCAGAGATTAAGCC	1
			Rev: Uni18S_R: ACGGAAACCTTGTTACGA	
Trematode universal 28 S	LSU (28 S)	MS	Fwd: LSU-5: TAGGTCGACCCGCTGAAYTTAAGCA	2
			Rev: 1500 R: GCTATCCTGAGGGAAACTTCG	
Trematode universal COI	cox1	MS/ MI	Fwd: JB3: TTTTTTGGGCATCCTGAGGTTTAT	3
			Rev: JB4.5: TAAAGAAAGAACATAATGAAAATG	
Family Heterophyidae	ITS2	MS/MI	Fwd: OPHET_Fwd: CTCGGCTCGTGTGTCGATGA	4
			Rev: OPHET_Rev: GCATGCARTTCAGCGGGTA	
Family Heterophyidae	LSU (28 S)	MI	Fwd: 28 S-Het-RFLP_F	5
			CTAACAAGGATTCCCTYAGTAAC	
			Rev: 28 S-Het-RFLP_R	
			TTCGATTAGTCTTTCGCCC	

(Continued)

Table 1. (Continued)

Parasite	DNA partition	Application	Primers (5′ → 3′)	References
Clinostomum spp.	*cox*1	MS/ MI	**Fwd:** 527 F: ATTCG(R)TTAAAT(Y) TKTGTGA	6
			Rev: 528 R: CCAAACYAACACMGACAT	
Clinostomum spp.	ITS	MS/ MI	**Fwd:** BD1: GTCGTAACAAGGTTTCCGTA	7
			Rev: BD2: ATCTAGACCGGACTAGGCTGTG	
Echinostoma spp.	*nad*1	MS/ MI	**Fwd:** NDJ11: AGATTCGTAAGGGGCCTAATA	8
			Rev: NDJ2a: CTTCAGCCTCAGCATAAT	
Echinostoma spp.	ITS	MS/ MI	**Fwd:** BD1: GTCGTAACAAGGTTTCCGTA	9
			Rev: BD2: ATCTAGACCGGACTAGGCTGTG	

Note: Fwd: Forward primer; Rev: Reverse primer; MS: Molecular Systematics; MI: Molecular Identification

REFERENCES

1. Dzikowski R, Levy MG, Poore MF, et al. Use of rDNA polymorphism for identification of Heterophyidae infecting freshwater fishes. *Dis Aqua Org* 2004;**59**:35–41.

2. Olsen PD, Cribb TH, Tkach VV, et al. Phylogeny and classification of Digenean (Plathyhelminthes: Trematoda). *Int J Parasitol* 2003;**33**:733–55.

3. Bowles J, McManus DP. Genetic characterization of the Asian Taenia, a newly described taeniid Cestode of huuman. *Am J Trop Med Hyg* 1994;**50**:33–44.

4. Skov J, Kania PW, Dalsgaard A, et al. Life cycle stage of Heterophyid trematodes in Vietnamese freshwater fishes traced by molecular and morphometric methods. *Vet Parasitol* 2009;**160**:66–75.

5. Thaenkham U, Phuphisut O, Pakdee W, et al. Rapad and simple identification of human pathogenic heterophyid intestinal fluke metacercariae by PCR-RFLP. *Parasitol Int* 2011;**60**:503–6.

6. Serono-Uribe A, Pinacho-Pinacho CD, García-Varela M, et al. Using mitochondrial and ribosomal DNA sequences to test the taxonomic validity of Clinostomum complanatum Rudolphi, 1814 in fish-eating birds and freshwater fishes in Mexico, with the description of the new species. *Parasiol Res* 2013;**112**:2855–70.

7. Luton K, Walker D, Blair D. Comparison of ribosomal internal transcribed spacer from two congeneric species of flukes (Plathyhelminthes: Trematoda: Digenea). *Mol Biochem Parasitol* 1992;**56**:323–8.

8. Morgan JAT, Blair D. Mitochondrial ND1 sequences used to identify echinostome isolates from Australia and New Zealand. *Int J Parasitol* 1998;**28**:493–502.

9. Morgan JAT, Blair D. Nuclear rDNA ITS sequence variation in the trematode genus Echinostoma: an aid to establishing relationships within the 37-collar-spine group. *Parasitol* 1995;**111**:609–15.

Biogeography The study of the distribution of species and ecosystems in geographic space and through geological time.

Cladistics An approach to biological classification in which organisms are grouped together based on whether or not they have one or more shared unique characteristics that come from the group's last common ancestor and are not present in more distant ancestors.

Congeneric species Species belonging to the same genus.

Cryptic species complex A group of organisms that are typically very closely related, yet their precise classifications and relationships cannot be easily determined.

Discrete data It is counted data. In phylogenetic analysis, discrete data usually consider qualitative data, including present/absent data and DNA sequence data.

Distance-matrix method The method of phylogenetic analysis explicitly relying on a measure of "genetic distance" between the sequences being classified.

Genetic marker A gene or DNA sequence with a known location on a chromosome that can be used to identify individuals or species.

Homoplasy Correspondence between parts or organs arising from evolutionary convergence.

Intraspecific Arising or occurring within a species; involving the members of one species.

Microsatellite DNA Also known as Simple Sequence Repeats (SSRs) or Short Tandem Repeats (STRs), they are repeating sequences of 2−6 base pairs of DNA.

Monophyly Groups of taxon (group of organisms) which forms a clade, meaning that it consists of an ancestral species and all its descendants.

Multigene family Groups of genes from the same organism that encode proteins with similar sequences either over their full lengths or limited to a specific domain.

Paraphyly Groups of taxon (group of organisms) which have a common ancestor but that do not include all its descendants.

PCR-RFLP Restriction fragment length polymorphism is a technique that exploits variations in homologous DNA sequences. It refers to differences between samples of homologous DNA molecules that come from differing locations of restriction enzyme sites and to a related laboratory technique by which these segments can be illustrated.

Phylogenetics The study of evolutionary relationships among groups of organisms, which are discovered through molecular DNA sequence data and morphological data matrices.

Phylogenetically informative site Given a set of homologous DNA sequences from four species, the positions where two species share a particular nucleotide and the other two species share a different nucleotide is called phylogenetically informative.

Phylogenomic The term has been used in multiple ways to refer to analysis that involves genome data and evolutionary reconstructions.

Phylogeography The study of the historical processes that may be responsible for the contemporary geographic distribution of individuals.

Population genetics The study of allele frequency distribution and change under the influence of the four main evolutionary processes: natural selection, genetic drift, mutation, and gene flow.

Phenotypic plasticity The ability of an organism to change its phenotype in response to changes in the environment.

RAPD RAPD is Random Amplified Polymorphic DNA. It is a type of PCR reaction in which the segments of DNA are amplified randomly by short primers (8−12 nucleotides).

Printed and bound by CPI Group (UK) Ltd, Croydon, CR0 4YY

03/10/2024

01040423-0008